Communications in Computer and Information Science **930**

Commenced Publication in 2007
Founding and Former Series Editors:
Phoebe Chen, Alfredo Cuzzocrea, Xiaoyong Du, Orhun Kara, Ting Liu,
Dominik Ślęzak, and Xiaokang Yang

More information about this series at http://www.springer.com/series/7899

Dmitry Ustalov · Andrey Filchenkov
Lidia Pivovarova · Jan Žižka (Eds.)

Artificial Intelligence and Natural Language

7th International Conference, AINL 2018
St. Petersburg, Russia, October 17–19, 2018
Proceedings

 Springer

Editors
Dmitry Ustalov
Data and Web Science Group
University of Mannheim
Mannheim, Baden-Württemberg
Germany

Andrey Filchenkov
ITMO University
St. Petersburg
Russia

Lidia Pivovarova
University of Helsinki
Helsinki
Finland

Jan Žižka
Mendel University
Brno
Czech Republic

ISSN 1865-0929   ISSN 1865-0937 (electronic)
Communications in Computer and Information Science
ISBN 978-3-030-01203-8   ISBN 978-3-030-01204-5 (eBook)
https://doi.org/10.1007/978-3-030-01204-5

Library of Congress Control Number: 2018955420

This Springer imprint is published by the registered company Springer Nature Switzerland AG
The registered company address is: Gewerbestrasse 11, 6330 Cham, Switzerland

Preface

The 7th Conference on Artificial Intelligence and Natural Language Conference (AINL), held during October 17–19, 2018, in Saint Petersburg, Russia, was organized by the NLP Seminar, ITMO University, and NLPub. Its aim was to (a) bring together experts in the areas of natural language processing, speech technologies, dialogue systems, information retrieval, machine learning, artificial intelligence, and robotics and (b) to create a platform for sharing experience, extending contacts, and searching for possible collaboration. The conference gathered more than 100 participants.

The review process was challenging. Overall, 56 papers were sent to the conference and only 19 were selected, for an acceptance rate of 34%. In all, 76 researchers from different domains were engaged in the double-blind reviewing process. Each paper received at least three reviews, in some cases, there were four reviews.

Altogether, 19 papers were presented at the conference, covering a wide range of topics, including morphology and word-level semantics, sentence and discourse representations, corpus linguistics, language resources, and social interaction analysis. Most of the presented papers were devoted to analyzing human communication and creating algorithms to perform such analysis. In addition, the conference program featured several special talks and events, including a plenary talk on societal challenges for information retrieval by Prof. Benno Stein, a tutorial on automatic text summarization by Dr. Sanja Štajner, a tutorial on creating virtual assistant skills in Just AI DSL by Darya Serdyuk and Svetlana Volskaya, industry talks and demos, and a poster session.

Many thanks to everybody who submitted papers and gave wonderful talks, and to those who came and participated without publication.

We are indebted to our Program Committee members for their detailed and insightful reviews; we received very positive feedback from our authors, even from those whose submissions were rejected.

We are grateful to our sponsors, Just AI, Huawei, and STC Group, for their support.

And last but not the least, we are grateful to our organization team: Anastasia Bodrova, Irina Krylova, Aleksandr Bugrovsky, Kseniya Buraya, Natalia Khanzhina, and Talgat Galimzhanov.

October 2018

Dmitry Ustalov
Andrey Filchenkov
Lidia Pivovarova
Jan Žižka

Organization

Program Committee

Mikhail Alexandrov	Autonomous University of Barcelona, Spain
Svetlana Alexeeva	Saint Petersburg State University, Russia
Artem Andreev	Institute for Linguistic Studies, Russian Academy of Sciences, Russia
Artur Azarov	Saint Petersburg Institute for Informatics and Automation, Russia
Rohit Babbar	Aalto University, Finland
Amir Bakarov	National Research University Higher School of Economics, Russia
Oleg Basov	Orel State University named after I.S. Turgenev, Russia
Erind Bedalli	University of Elbasan "Aleksandër Xhuvani", Albania
Anton Belyy	ITMO University, Russia
Siddhartha Bhattacharyya	RCC Institute of Information Technology, India
Chris Biemann	Universität Hamburg, Germany
Elena Bolshakova	Lomonosov Moscow State University, Russia
Pavel Braslavski	Ural Federal University, Russia
Maxim Buzdalov	ITMO University, Russia
John Cardiff	ITT Dublin, Ireland
Dmitry Chalyy	Yaroslavl State University, Russia
Mikhail Chernoskutov	Krasovskii Institute of Mathematics and Mechanics, Russia
Bonaventura Coppola	University of Trento, Italy
Frantisek Darena	Mendel University Brno, Czech Republic
Boris Dobrov	Lomonosov Moscow State University, Russia
Ekaterina Enikeeva	Saint Petersburg State University, Russia
Vera Evdokimova	Saint Petersburg State University, Russia
Elena Filatova	City University of New York, USA
Andrey Filchenkov	ITMO University, Russia
Tommaso Fornaciari	Bocconi University in Milan, Italy
Natalia Grabar	CNRS, Université de Lille, France
Jiri Hroza	GAUSS Algorithmic, Czech Republic
Dmitry Ignatov	National Research University Higher School of Economics, Russia
Vladimir Ivanov	Innopolis University, Russia
Alexander Jung	Aalto University, Finland
Nikolay Karpov	National Research University Higher School of Economics, Russia
Egor Kashkin	V.V. Vinogradov Russian Language Institute, Russia

Denis Kirjanov	Sberbank of Russia, Russia
Pavel Klinov	Stardog Union, USA
Daniil Kocharov	Saint Petersburg State University, Russia
Yury Kochetov	Sobolev Institute of Mathematics, Russia
Maxim Kolchin	ITMO University, Russia
Mikhail Korobov	ScrapingHub Inc., Russia
Evgeny Kotelnikov	Vyatka State University, Russia
Tomas Krilavičius	Baltic Institute of Advanced Technology, Lithuania and Vytautas Magnus University, Lithuania
Andrey Kutuzov	University of Oslo, Norway
Natalia Loukachevitch	Lomonosov Moscow State University, Russia
Artem Lukanin	European Patent Office, The Netherlands
Alexey Malafeev	National Research University Higher School of Economics, Russia
Vladislav Maraev	University of Gothenburg, Sweden
Roman Meshcheryakov	Tomsk State University of Control Systems and Radioelectronics, Russia
George Mikros	National and Kapodistrian University of Athens, Greece
Tristan Miller	Technische Universität Darmstadt, Germany
Alexander Molchanov	PROMT, Russia
Kirill Nikolaev	National Research University Higher School of Economics, Russia
Allan Payne	SORC, UK
Georgios Petasis	National Centre of Scientific Research "Demokritos", Greece
Stefan Pickl	Bundeswehr University Munich, Germany
Lidia Pivovarova	University of Helsinki, Finland
Vladimir Pleshko	RCO, Russia
Alexey Romanov	University of Massachusetts Lowell, USA
Paolo Rosso	Universitat Politècnica de València, Spain
Yuliya Rubtsova	A.P. Ershov Institute of Informatics Systems, Russia
Eugen Ruppert	Universität Hamburg, Germany
Ivan Samborskii	National University of Singapore
Andrey Savchenko	National Research University Higher School of Economics, Russia
Christin Seifert	University of Twente, The Netherlands
Alexander Semenov	University of Jyvaskyla, Finland
Alexey Sorokin	Lomonosov Moscow State University, Russia and Moscow Institute of Science and Technology, Russia
Aleksandr Tarelkin	EPAM, Russia
Irina Temnikova	Qatar Computing Research Institute, Qatar
Mike Thelwall	University of Wolverhampton, UK
Elena Tutubalina	Kazan (Volga Region) Federal University, Russia
Dmitry Ustalov	University of Mannheim, Germany
Elior Vila	University of Elbasan "Aleksandër Xhuvani", Albania
Roman Yangarber	University of Helsinki, Finland

Wajdi Zaghouani	Hamad Bin Khalifa University, Qatar
Marcos Zampieri	University of Wolverhampton, UK
Alexey Zobnin	National Research University Higher School of Economics, Russia
Nikolai Zolotykh	University of Nizhni Novgorod, Russia
Jan Žižka	Mendel University in Brno, Czech Republic

Contents

Language Resources

Social Interaction Analysis

Morphology and Word-Level Semantics

Deep Convolutional Networks for Supervised Morpheme Segmentation of Russian Language

Alexey Sorokin[1,2] and Anastasia Kravtsova[1(✉)]

[1] Faculty of Mechanics and Mathematics, Lomonosov Moscow State University,
Moscow, Russia
alexey.sorokin@list.ru, nastik_pretty@mail.ru
[2] Moscow Institute of Physics and Technology, Moscow, Russia

Abstract. The present paper addresses the task of morphological segmentation for Russian language. We show that deep convolutional neural networks solve this problem with F1-score of 98% over morpheme boundaries and beat existing non-neural approaches.

Keywords: Morpheme segmentation · Neural networks · Evaluation

Many successful approaches of modern NLP treat words as mere sequences of symbols, not as atomic units, for example, the FastText model constructs word embedding from the embeddings of its ngrams. However, not all symbol ngrams are of the same utility, the most important ones often correspond to morphemes or pseudomorphs: the root is essential in capturing word semantics, while affixes reflect morphological and syntactic relations. It brings the problem of automatic morphological segmentation to the fore of computational linguistics. Recently, morpheme segmentation was used as a part of machine translation system in [9] and in [1] for constructing word embeddings.

In earlier years of NLP this task was usually solved using no or minimal supervision. Researchers tried to utilize letter variety statistics [3], or modeled the sequence of morphological segments using either HMM [2] or adaptor grammars [6]. Being linguistically motivated, this approach obviously suffers from its unsupervised nature and predetermined constraints imposed by the probabilistic model. As any segmentation task, morpheme segmentation can be transformed to a sequence tagging problem using BMES-scheme. However, for most languages the amount of supervised data is too low for such treatment. Fortunately, Russian does not have this problem since the morphological dictionary of Tikhonov [8] which includes more than 90000 lexemes is freely available.

As most sequence tagging tasks, morphological segmentation was successfully addressed using conditional random fields [4]. For other sequence labelling tasks,

The work is partially supported by National Technological Initiative and Sberbank, project identifier 0000000007417F630002.

D. Ustalov et al. (Eds.): AINL 2018, CCIS 930, pp. 3–10, 2018.
https://doi.org/10.1007/978-3-030-01204-5_1

such as NER or morphological tagging, neural network approaches tend to outperform CRFs. Therefore our choice is to apply neural networks to morpheme segmentation. Since morphological segmentation is essentially local—the boundary position depends mostly on the immediate context—we apply convolutional neural networks (CNNs) due to their excellent ability to capture local phenomena. Neural networks were applied to morpheme segmentation [5,7], however, we do not know any works testing CNNs for this problem.

Our contribution is threefold: we release a cleared version of A. N. Tikhonov morphological dictionary, we test a multilayer convolutional network as a baseline approach for this task and show its effectiveness; also we demonstrate that additional memorizing of morphemes slightly improves performance. All the data and code are available on Github[1].

1   Model Architecture

We treat morpheme segmentation as a sequence labeling task. Segments are encoded using BMES-scheme, where B stands for Begin, M—for Middle and E and S for End and Single respectively. As in named entity recognition, we also encode types of the morphemes to be tagged, for example, the encoding of the word *учитель* (*teacher*) and its segmentation *уч/ROOT/и/SUFF/тель/SUFF/* is

у	ч	и	т	е	л	ь
B-ROOT	E-ROOT	S-SUFF	B-SUFF	M-SUFF	M-SUFF	E-SUFF

Our network starts from 0/1 encodings of the input letters. These vectors are passed through several convolutional layers. For each window width we have its own filters to deal with symbol ngrams of different length. The outputs of all convolutions are concatenated and passed through a (possibly) multilayer perceptron. The final layer of the perceptron is followed by softmax which outputs probability distribution over possible labels in each position of the word.

Describing the model formally, for a word $w = w_1 \ldots w_n$ and one-hot encoding of its letters $e_1 \ldots e_n$, we have

$$(z_1^1)_i = \mathrm{CONV}(e_{i-d_1/2}, \ldots, e_i, \ldots, e_{i+d_1/2}), i = 1, \ldots, n$$
$$\ldots$$
$$(z_r^1)_i = \mathrm{CONV}(e_{i-d_r/2}, \ldots, e_i, \ldots, e_{i+d_r/2}),$$
$$\ldots$$
$$(z_j^k)_i = \mathrm{CONV}((z_j^{k-1})_{i-d_1/2}, \ldots, (z_j^{k-1})_i, \ldots, (z_j^{k-1})_{i+d_1/2}),$$

Here $k = 2, \ldots, K$ is the number of layer, $d_1, \ldots, d_r$ are different window widths and i is position in the sequence. For each i we concatenate all convolutions as $z_i = [(z_1^K)_i, \ldots, (z_r^K)_i]$. z_i encodes the context around i-th position and is further passed through a two layer perceptron that outputs a probability distribution p_i over all possible classes:

---

[1] https://github.com/AlexeySorokin/NeuralMorphemeSegmentation.

$$h_i' = \max{(W'z_i + b', 0)},$$
$$h_i = Wh_i + b$$
$$p_{ij} = \frac{e^{h_{ij}}}{\sum\limits_{r} e^{h_{ir}}}$$

Weights of all the layers are optimized during model training.

2 Experiments

2.1 Data

We used the electronic version of Tikhonov dictionary available in the Web. Since downstream tasks often require morpheme types (affixes for morphological tasks and roots for semantic ones), we labeled the morphemes using the data from slovolit.ru. This data was manually postprocessed and cleared to avoid errors and inconsistencies.

We used 7 types of morphemes: prefix, root, suffix, ending, postfix (-ся in целоваться), link (средн-е-русский) and hyphen (англо-русский). Data was partitioned in 3/1 proportion and the same partition was held for all the experiments. Training part contained 72033 words and the test part includes 24012 ones. All the data is available on our Github page.

2.2 Model Implementation

We implement our model using Keras library with Tensorflow backend. 20% of the training data was left for validation. We stopped learning when accuracy on this development set did not improve for 10 epochs and trained the model for maximum of 75 epochs. The size of batch was 32. We used standard Adam optimizer with default parameters except for gradient clipping whose threshold was set to 5.0. In addition to the architecture described in the previous section we used ReLU activations on all the layers and inserted a dropout layer with dropout rate 0.2 between consecutive convolutional layers.

2.3 Experiments

We tested different parameters of our neural network: the number of convolutional layers (1, 2 or 3) and the distribution of filters on each layer. In preliminary experiments we selected optimal total number of filters of 192 and further tried different combinations of window sizes. We evaluated 4 combinations: 64 filters of width 3, 5, 7, 96 filters of width 5 and 7 and 192 filters either of width 5 or 7. For 3 layers we tested only two best combinations from previous tests. We report 5 evaluation measures: the usual precision, recall and F1-measure over morpheme boundaries, accuracy of BMES labels and percentage of correctly segmented words. To be ranked as true, not only the position of the morpheme, but also its type should coincide with the correct one. In contrast to [4], we take

Table 1. Quality of morpheme segmentation.

# layers	Filter combination	Precision	Recall	F1-score	Accuracy	Word accuracy
1	64-64-64	96.14	95.63	95.88	92.74	76.21
	96-96	96.31	*95.90*	96.10	93.12	77.47
	192(5)	96.17	95.38	95.77	92.59	75.82
	192(7)	*96.49*	95.73	*96.11*	*93.17*	*77.62*
2	64-64-64	97.12	97.2	97.16	94.93	83.13
	96-96	97.18	97.41	97.29	95.16	83.80
	192(5)	*97.18*	*97.59*	*97.38*	*95.34*	*84.48*
	192(7)	96.79	97.40	97.09	94.85	82.69
3	96-96	97.47	97.70	97.58	95.76	85.67
	192(5)	**97.67**	**97.82**	**97.74**	**95.99**	**86.42**

the final word boundary into account since we may fail to predict it properly by choosing an incorrect morpheme type.

Results of evaluation are shown in Table 1, where the leftmost column stands for the number of layers and the next one for the combination of filters. For a single layer width 7 is optimal since 5 symbols are too little to capture long roots. With 2 and 3 layers width 5 is better since two layers of width 5 collect information from 9 consecutive symbols. Using 2 layers instead of 1 leads to error reduction over 30% for all the metrics, the improvement between 2 and 3 layers is much less but also clear. We tested a bidirectional LSTM instead of final convolutional layer but it deteriorated performance. That shows that morphological segmentation is essentially a local task and does not require memorizing long-distance dependencies.

We suggest that convolutional filters have enough power to learn morphs, but also experimented with memorizing morphemes directly. The context of each position is encoded with a 15-dimensional vector. This vector contains three boolean values for each principal morpheme type (prefix, root, suffix, ending and postfix): whether there is a morpheme ngram that begins in this position, ends in it or whether current letter can be a single-letter morpheme. The vector is concatenated with symbol encoding. We extract all morphemes that occur at least 3 times in the training set. We tested two variants of memorization: the basic one using 0/1 features and the one equipped with ngram counts: if an ngram *чит* was labeled as root 5 times and appears 10 times in corpus, it will get score 0.5 for root features. For prefixes we use only ngrams in the beginning of the word, for endings and postfixes – only in the end. We observed that there is no difference between performance of these two models. Table 2 shows that memorizing is beneficial for a single layer model, while for 3 layers the improvement is only marginal. It proves that deep convolutional architecture has enough power to learn useful symbol ngrams.

For the top score we ensembled three models with different random initializations and averaged the probabilities they predict. As shown in Table 3, that improved performance additionally; note that averaging has higher effect than memorization.

Table 2. Effect of memorization on morpheme segmentation for different number of layers.

# layers	Filter combination	Precision	Recall	F1-score	Accuracy	Word accuracy
1	192(7)	96.49	95.73	96.11	93.17	77.62
	192(7) + memo	95.96	97.09	96.52	93.91	80.14
2	192(5)	97.18	97.59	97.38	95.34	84.48
	192(5) + memo	97.22	97.80	97.51	95.61	85.29
3	192(5)	97.67	97.82	97.74	95.99	86.42
	192(5) + memo	97.67	97.97	97.82	96.15	87.03

Table 3. Effect of ensembling and memorization on morpheme segmentation for best model configuration.

# layers	Filter combination	Precision	Recall	F1-score	Accuracy	Word accuracy
3	192(5)	97.67	97.82	97.74	95.99	86.42
	192(5) + ensemble	97.99	98.00	98.00	96.45	87.99
	192(5) + memo	97.67	97.97	97.82	96.15	87.03
	192(5) + memo + ensemble	97.86	98.35	98.10	96.64	88.62

Since usually morpheme segmentation is done in low-resource setting, we evaluated our model on different amounts of training data. The learning curves are presented on Fig. 1. Already 20% of the training data (about 14000 words) are enough to outperform CRF-based model of [4] (see comparison below).

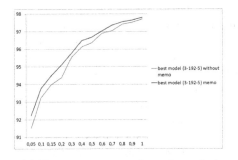
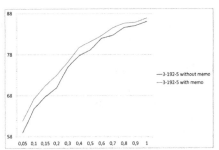

(a) F1-score over morpheme boundaries   (b) Word accuracy

Fig. 1. Dependence of quality from training data fraction

We compared our model against CRF-based state-of-the-art semi-supervised system of [4]. We used our own evaluation script and report F1-score over morpheme boundaries in two variants:

1. Without evaluating morpheme types and taking final boundary into account, as it is usually reported.
2. With morpheme types for our system and without morpheme types for the model of [4] since it does not use and output them.

We also report word-level accuracy with and without morpheme types (Table 4).

Table 4. Comparison with [4]

Metrics	[Ruokolainen]	[Ruokolainen, Harris features]	Our model (+memo)
Morph boundary F1 (without last boundary and types)	92.17	92.95	97.16
Morph boundary F1 (with last boundary and types)	94.24	94.77	97.9
Word accuracy (without types)	65.29	68.19	87.53
Word accuracy (with types)	65.29	68.19	87.03

In addition to its lower accuracy, even the basic model from [4] trained for more than 1 h on a single core Intel Xeon CPU with 512 GB RAM[2] and ran out of memory on 6 GB RAM laptop, while our model takes about 30 min to be trained on 2 GB GPU on the same laptop. Actually, a neural network has about 10 times fewer parameters than CRF, which explains its lower requirements.

2.4 Error Analysis

Though our system sets a new benchmark for Russian morpheme segmentation, it still has much space to improve. Actually, for more than 10% of words the predicted morpheme structure is incorrect. We present a short analysis of our topmost model (ensemble of 3 models with 3 layers and 192 neurons on each, window width 5) errors and start with the distribution of error types, given in Table 5. Note a pleasant fact, that only in 1.67% of cases (402 of 24012) our model fails for more than 1 morpheme boundary.

We also inspect the dependency of model performance from the number of morphemes. The results (see Table 6) are quite surprising: the lowest quality is observed for 1 or 2 morphemes in the word. The model typically overgenerates, some examples of its errors are given in Table 7. Often our system incorrectly selects word segments which are homonymical to frequent morphemes, while in other cases it reconstructs etymologically correct morphemes which have either desemantized in synchrony or are not labeled due to annotation conventions. A minor part of such mistakes may be due to errors in available electronic version of Tikhonov dictionary (see 7b).

---

[2] The model equipped with Harris features takes more than 2 h.

Table 5. Distribution of error types

# bound errors	Error type	Count	Percentage
0	Correct	21280	88.62
	Morpheme type	109	0.45
1	Overgeneration	1107	4.61
	Undergeneration	893	3.72
	Bound position	221	0.92
≥2	Overgeneration	192	0.81
	Undergeneration	99	0.41
	Other	111	0.46

Table 6. Performance quality depending on the number of morphemes

# Morphemes	1	2	3	4	5	≥6
Word accuracy	83.27	78.75	85.24	92.40	93.30	86.33
# in training set	3710	7599	15983	22472	16380	5890

Table 7. Examples of annotation errors.

(a) Segmentation errors caused by homonymy.

Correct	Predicted
недосуг	не-до-суг
облучок	об-луч-ок
помост	по-мост
аполог-ет	апол-о-гет
буханк-а	бух-ан-к-а

(b) Segmentation errors caused by desemantization or annotation errors.

Correct	Predicted
окоп-ник	о-коп-ник
поступательн-ый	поступ-а-тельн-ый
очевидн-ый	очевид-н-ый
оклад	о-клад
ловк-ий	лов-к-ий

3 Conclusions and Future Work

We have developed a convolutional tagging model for Russian morpheme segmentation task. It clearly outperforms all earlier approaches consuming less computational resources. Further directions are multifold: the first one is to extend the model to operate in semi-supervised fashion with very little data available. Another task to address is reconstruction of deep morpheme structure which requires allomorphy reduction (e.g., Russian *c-* and *co-* or English *-s* and *-es* should be mapped to the same morpheme) as in [5]. The third task is to test whether obtained morpheme segmentations are useful in downstream morphological or semantic tasks such as morphological tagging or checking semantic relatedness.

References

1. Botha, J., Blunsom, P.: Compositional morphology for word representations and language modelling. In: International Conference on Machine Learning, pp. 1899–1907 (2014)
2. Creutz, M., Lagus, K.: Unsupervised morpheme segmentation and morphology induction from text corpora using Morfessor 1.0. Helsinki University of Technology, Helsinki (2005)
3. Harris, Z.S.: Morpheme boundaries within words: report on a computer test. In: Harris, Z.S. (ed.) Papers in Structural and Transformational Linguistics. FLIS, pp. 68–77. Springer, Dordrecht (1970). https://doi.org/10.1007/978-94-017-6059-1_3
4. Ruokolainen, T., et al.: Painless semi-supervised morphological segmentation using conditional random fields. In: Proceedings of the 14th Conference of the European Chapter of the Association for Computational Linguistics, vol. 2: Short Papers, pp. 84–89 (2014)
5. Ruzsics, T., Samardzic, T.: Neural sequence-to-sequence learning of internal word structure. In: Proceedings of the 21st Conference on Computational Natural Language Learning (CoNLL 2017), pp. 184–194 (2017)
6. Sirts, K., Goldwater, S.: Minimally-supervised morphological segmentation using adaptor grammars. Trans. Assoc. Comput. Linguist. **T. 1**, 255–266 (2013)
7. Shao, Y.: Cross-lingual word segmentation and morpheme segmentation as sequence labelling. arXiv preprint arXiv:1709.03756 (2017)
8. Tikhonov, A.N.: Morphemno-orfograficheskij slovar, 704 c. ACT Publishing (2002). (in Russian)
9. Vylomova, E., et al.: Word representation models for morphologically rich languages in neural machine translation. arXiv preprint arXiv:1606.04217 (2016)

Smart Context Generation
for Disambiguation to Wikipedia

Andrey Sysoev[1(✉)] and Irina Nikishina[1,2]

[1] Ivannikov Institute for System Programming, Russian Academy of Sciences,
Moscow, Russia
{sysoev,nia}@ispras.ru
[2] Higher School of Economics, Moscow, Russia

Abstract. Wikification is a crucial NLP task that aims to identify entities in text and disambiguate their meaning. Being partially solved for English, the problem still remains fairly untouched for Russian. In this article we present a novel approach to Disambiguation to Wikipedia applied to the Russian language. Inspired by the Neural Machine Translation task our method implements encoder-decoder neural network architecture. It translates text tokens into concept embeddings that are subsequently used as context for disambiguation. In order to test our hypothesis we add our context features to GLOW system considered a baseline. Moreover, we present commonly available dataset for the Disambiguation to Wikipedia task.

Keywords: Disambiguation to Wikipedia · Wikification for Russian
Encoder-decoder neural network architecture · Concept embeddings

1 Introduction

It is widely acknowledged that Wikipedia has almost become the most popular and authoritative source in the modern Internet society, remaining the largest multi-language corpus that is especially useful for different NLP tasks. In particular, Wikipedia might be useful for Named Entity recognition [17,20], word sense disambiguation [5], text classification [13] and other tasks that require additional information about real world enitites that could be gained by means of Wikification.

Wikification task consists of two levels: one is responsible for locating entities in raw text, the other stands for associating entities with the appropriate Wikipedia pages – hereinafter concepts. The last step is also known as Disambiguation to Wikipedia (D2W) and might also be considered a separate task: to each mention m assign a Wikipedia concept e or a special *nil* value (not-yet-in-Wikipedia concept). For instance, "St. Petersburg" in sentence "First time I saw St. Petersburg last year" may refer either to the Russian city or to the city in the United States or even to the Iranian comedy film. The goal of a D2W

© Springer Nature Switzerland AG 2018
D. Ustalov et al. (Eds.): AINL 2018, CCIS 930, pp. 11–22, 2018.
https://doi.org/10.1007/978-3-030-01204-5_2

system in this case is to associate "St. Petersburg" with the correct Wikipedia concept.

While most papers about Wikification and D2W describe new methods implemented for English or other European languages, very little research is made for Russian. That is why in the current paper we present a novel approach to D2W in application to the Russian language.

We assume that context used for disambiguation may also be generated with the help of Neural Machine Translation (NMT) techniques. Thus our idea is to build a system that transforms text tokens into a set of concept embeddings – smart context.

According to our hypothesis of "token-to-concept" translation, sentence "She ate too much Caesar at Gordon Ramsay yesterday" should be translated to the language of concepts as "Caesar salad, Restaurant Gordon Ramsay" and not "Julius Caesar, Gordon Ramsay". We expect concept embeddings generated by NMT model act as appropriate unambiguous context for the D2W task.

In order to evaluate usefulness of proposed features, based on similarity to generated smart context, we implement the approach from [18] as the baseline. We also create a dataset for the Russian language, which is described in Sect. 5.1.

Therefore, the main contribution of our research is the following: we apply the existing D2W method to Russian, demonstrate the advantages of the developed smart context based features and propose the generated dataset as the gold standard dataset for the D2W task.

2   Related Work

Our approach is based on application of encoder-decoder architecture borrowed from NMT research area to solve D2W problem. That is why we suppose being important to review related work in both fields.

2.1   Wikification and Disambiguation

As a subtask of Entity Linking, Wikification for the English language has quite a long history. The whole timeline is perfectly described in [22], while we draw our attention to those works which are more important for our research.

First, two prominent studies for Wikification and D2W tasks should be mentioned: [16] where standard measures like commonness and relatedness are proposed and [18] that introduces Global and Local algorithms for Entity Disambiguation (GLOW). GLOW system from the second paper is also described in Sect. 3 in more detail.

Furthermore, we should mention [6] as it proves Entity Linking to be quite useful for other NLP tasks. In [7] it is also demonstrated that capturing topic at multiple granularities from text via a CNN model is essential for concept disambiguation.

Besides our research the idea of generating concept embeddings is also developed in [8]. Concept vectors are trained there using word2vec [15] and then utilized for generating local context attention. For global disambiguation they propose using Conditional Random Fields and Loopy Belief Propagation. The authors compare their approach to and mostly outperform [3] and [11].

One of the most recent works about D2W is [23], in which authors apply the Random Forest algorithm for mention disambiguation. To decide whether an entity should be included to the result set they use helpfulness evaluation based on link probability, entity popularity, entity class and topical coherence.

Concerning D2W for the Russian language, we suppose that its current state is only a starting point. Besides [21] who try to implement maximum entropy classifier likewise in [16] for Russian and test in on private corpus, we are not aware of other works devoted to the current topic.

2.2 Neural Machine Translation

With the recent developments in Deep Neural Networks, NMT is closely associated with sequence-to-sequence model [19]. This approach generally comprises two stages: *encoding stage* that converts sentence from source language into a vector representing its language-independent meaning and *decoding stage*, responsible for translating this vector into sentence written in target language.

A few years ago NMT systems like [4] and [24] implemented bidirectional Long Short-Term Memory (LSTM) models for both encoding and decoding phases. Later [9] integrated Convolutional Neural Networks (CNN), applying convolutional model instead of LSTM to encoder and then even to decoder [10]. In the current study we are comparing biLSTM and CNN based encoders (Sects. 4.2 and 4.3) with regards to the D2W task.

Another constituent part of NMT model is the attention mechanism that allows to learn alignments between source and target sentences. For the first time it is used in [1], then improved by [14]. Application of attention weights for the current research might be rather unevident and is thoroughly described in Sect. 4.4.

3   Baseline

As a baseline solution for D2W task we select GLOW approach from [18]. In this section we briefly describe the algorithm itself along with our modifications and clarifications.

GLOW starts with enriching provided collection $\{m_1, m_2, \ldots, m_N\}$ with extra mentions, computed as named entities and noun phrases of length not more than 5. Each mention m is associated with its possible meanings E_m, extracted from Wikipedia redirects and anchor texts; in correspondence to [18], only top 20 most frequent concepts are analysed.

Then come two main GLOW phases: first of all, global context, which consists of a number of input text describing concepts, is identified; secondly, this context is used to determine the final assignment of concepts to input mentions. Each phase is based on ranker-linker pair of machine learning algorithms which differ only in the set of features being used.

Ranker accepts a mention m with possible meanings E_m within a document D and grades all E_m according to their plausibility of being correct disambiguation of m. Ranker training is performed with RankSVM [12].

For each mention m **linker** is provided with ranker-computed scores of possible meanings; its goal is to filter out presumably incorrect assignments, when ranker fails to deliver the highest weight to the correct meaning. Linker is a conventional binary linear SVM classifier.

During context generation phase ranker uses context independent and local context features. For computing final meaning-concept assignment global context features are used as well.

3.1   Context Independent Features

Context independent features include $P(e|m)$ and $P(e)$. $P(e|m)$ – commonness – indicates how often mention m links to entity e in Wikipedia. $P(e)$ is the portion of Wikipedia articles, which have links to e.

3.2   Local Context Features

Let us introduce the following notations: $text(m)$ is TFIDF vector of document D, containing m; $context(m)$ is TFIDF vector computed for w-size window around m; $text(e)$ contains $2w$ elements with top TFIDF weight extracted from Wikipedia page corresponding to concept e; $context(e)$ is similar to $text(e)$ but is collected through all w-size windows around mentions, linked to e throughout the whole Wikipedia. In contrast to [18] we utilize lemmas instead of tokens to gain $text(\cdot)$ and $context(\cdot)$.

Local context features include cosine similarity calculated for the following pairs of vectors:

$text(m) \leftrightarrow text(e),$
$text(m) \leftrightarrow context(e),$
$context(m) \leftrightarrow text(e),$
$context(m) \leftrightarrow context(e).$

Additionally, [18] uses reweighted versions of described features, which are aimed at changing token importance in TFIDF vectors: boost more specific and fine less specific tokens for the given possible meanings of the mention m. Reweighted TFIDF is evaluated according to the formula:

$$w_{text}(l, e, m) = \frac{text(e)_{[l]}}{\sum_{e' \in E_m} text(e')_{[l]}}, \tag{1}$$

where $text(e)_{[l]}$ is weight of lemma l in TFIDF vector $text(e)$, which is assumed to be 0 if vector $text(e)$ does not contain l. $w_{context}(l, e, m)$ is computed similarly.

3.3   Global Context Features

Let us introduce some notations (see Table 1).

Table 1. Notations for computing global context features.

agg_G	*Aggregating function*
max_G	Maximum value computed throughout the whole global context G
avg_G	Average value computed throughout the whole global context G
$\mathbb{1}$	*Concepts link indicator*
$\mathbb{1}_{e_i - e_j}$	Binary indicator of e_i having a link to e_j *or* vice versa
$\mathbb{1}_{e_i \leftrightarrow e_j}$	Binary indicator of e_i having a link to e_j *and* vice versa
sim	*Similarity*
PMI'	Pointwise Mutual Information similarity measure (formulas 2 and 3)
NGD	Normalized Google Distance (formula 4)
$links$	*Link set*
in_links	Set of concepts, which have an outgoing link to e
out_links	Set of concepts, to which e has an outgoing link

$$PMI' = \frac{PMI}{1 + PMI}, \tag{2}$$

$$PMI(L_1, L_2) = \frac{|E||L_1 \cap L_2|}{|L_1||L_2|}, \tag{3}$$

$$NGD(L_1, L_2) = \frac{log\ max(|L_1|, |L_2|) - log|L_1 \cap L_2|}{log|E| - log\ min(|L_1|, |L_2|)}, \tag{4}$$

where E is a set of all Wikipedia concepts.

Global features are constructed by composing combinations of introduced options, peeking one at a time: $agg_{g \in G}\ \mathbb{1} \cdot sim(links(e), links(g))$. For instance, a sample feature $F(e)$ is $max_{g \in G}\ \mathbb{1}_{e-g} \cdot PMI'(in\_links(e), in\_links(g))$.

A pair of extra global features utilized in GLOW is $max_{g \in G}\ \mathbb{1}_{e \leftrightarrow g}$ and $avg_{g \in G}\ \mathbb{1}_{e \leftrightarrow g}$.

3.4   Linker Features

Linker features include the same set of features as its corresponding ranker. However, there is a number of additional features:

- difference in score between the best and the second-best concept, produced by ranker;
- entropy of possible mention meanings;
- indicator of meaning being a named entity;

- the fraction of mention appearances in Wikipedia, where it is used as a link;
- Good-Turing estimate of mention not having correct meaning described in Wikipedia. We use the following formula:

$$F_{GoodTuring}(m) = \frac{\sum_{e \in E_m} \mathbb{1}_{count(e,m)=1}}{\sum_{e \in E_m} count(e, m)}, \tag{5}$$

where $count(e, m)$ is the number of times mention m is linked to concept e in Wikipedia.

4   Similarity to Generated Context

In this section we introduce a novel type of context, which is exploited in D2W. Additionally, we propose a method for computing similarities from concept to generated context, which are used as extra features in GLOW algorithm.

4.1   Context Generation

The proposed type of context – smart context – is the result of translating input text into a "language of concepts" with some neural machine translation approach. In this work we utilize a simple encoder-decoder architecture, proposed in [19].

Input tokens (with special END_TOKEN appended) are mapped into their embeddings and then fed into encoder part of the network. Encoder translates them into internal representation I, which is further passed to decoder. Additionally, encoder compresses the whole input into a pair of fixed-length vectors (c, h), which are later used in decoder initialization (see Fig. 1).

Decoder part of neural network is based on LSTM. Encoder's (c, h) pair is passed through fully-connected layers F_c and F_h to match decoder LSTM state size and is used to initialize it. Tokens internal representation I is aggregated in conformance to attention mechanism [1] and further fed into decoder LSTM. Moreover, each LSTM cell also consumes decoder output from the previous step (initially, special START_CONCEPT is passed instead). Each LSTM cell output and newly computed attention vector are passed through fully-connected layer F_s, L2-normalized and then considered a target context concept embedding. Decoding stops when special END_CONCEPT embedding is produced.

We experiment with two types of encoder architectures – biLSTM-based and CNN-based, which are described in detail in Sects. 4.2 and 4.3 correspondingly.

4.2   BiLSTM-Based Encoder

Token embeddings are passed as input to biLSTM (each LSTM is of size l), which converts them into internal representation I. Hidden state vectors $\overrightarrow{c}$ and $\overleftarrow{c}$ of forward and backward LSTMs are concatenated to form final vector c. (h is computed in the same way). To introduce regularization to our model, dropout layers with keeping probability p are applied to I, c and h before returning them from encoder (see Fig. 2).

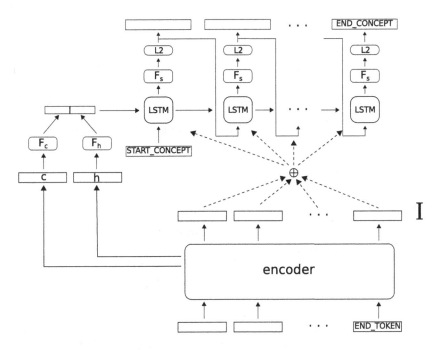

Fig. 1. Encoder-decoder neural network architecture.

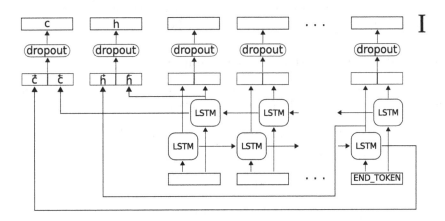

Fig. 2. BiLSTM-based encoder.

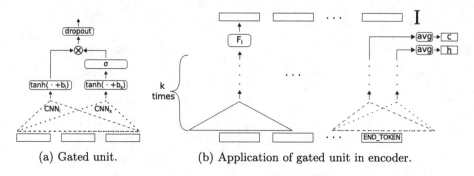

(a) Gated unit.   (b) Application of gated unit in encoder.

Fig. 3. CNN-based encoder.

4.3 CNN-Based Encoder

CNN-based encoder architecture is hugely inspired by [10]; it mainly consists of several CNN-based gated units (see Fig. 3a):

$$g(T) = v_l(T) \otimes \sigma(v_s(T)), \tag{6}$$

$$v.(T) = tanh(cnn.(T) + b.), \tag{7}$$

where T is a matrix of token embeddings, $\otimes$ is the pointwise vector multiplication, $cnn(T)$ is the result of application of CNN layer to matrix T, b is a trainable variable. Output of each unit is then passed through dropout layer with keeping probability p.

CNN-based gated units are stacked one upon another for k times. Output of the final block is averaged; it constitutes the final c vector. Vector h is computed in the same way, but using a separate stack of blocks. Another stack is used to compute internal representations I, but its final block output is preliminary traversed through fully-connected layer F_I (see Fig. 3b).

4.4 Similarity to Smart Context Computation

Similarity features $F_S(e, m)$ from concept e to smart context S for mention m are computed according to the following formulas:

$$F_S^{max}(e, m) = \max_{s \in S} cos(embedding(e), s), \tag{8}$$

$$F_S^{avg}(e, m) = \frac{1}{|S|} \sum_{s \in S} cos(embedding(e), s), \tag{9}$$

$$F_S^{attention_{max}}(e, m) = \max_{s \in S} \frac{\alpha(s, m)}{\sum_{s \in S} \alpha(s, m)} cos(embedding(e), s), \tag{10}$$

$$F_S^{attention_{avg}}(e, m) = \frac{1}{\sum_{s \in S} \alpha(s, m)} \sum_{s \in S} \alpha(s, m) cos(embedding(e), s), \tag{11}$$

$$\alpha(s, m) = \sum_{t:\ t\ intersects\ m} attention(t, s), \tag{12}$$

where t is text token, $attention(t, m)$ is decoder attention for token t when computing concept embedding s. In other words, to compute each context embedding weight $\alpha(s, m)$ we sum attention scores of mention tokens, returned by decoder.

5 Evaluation

In the current section we describe dataset prepared for training and testing GLOW and neural network parameters. Furthermore, we evaluate the results obtained from the algorithms described above.

5.1 Data and Parameters

While for the English language there exists a large amount of corpora for the disambiguation task, there is no open dataset available for Russian. Thus, we download the Russian Wikipedia dump of May 1, 2018 that contains more than 1470000 articles and build our own corpus. We collect those pages that attain one of the two best grades in WikiProject article quality evaluation scheme: we select 2968 articles from labelled as `Good article` for training; 1056 articles categorized as `Featured article` are treated as test set[1]. For training our neural network models we omit `Featured articles` in order to avoid possible overlapping with test data.

Moreover, the Wikipedia dump is utilized for fitting embedding models. We pre-train a word2vec [15] model (size = 100, window = 5, skip-gram) for generating concept embeddings and a fasttext [2] model (size = 100, window = 5, skip-gram) for tokens.[2]

Table 2 outlines neural network hyperparameter values used during the experiments.

5.2 Evaluation Results

In this section we explore the usefulness of our features, based on concept similarity to smart context, on the D2W task.

In order to carry out fair evaluation we calculate the minimum level for the results which is known as `Most common sense` (MCS in Table 3). For each mention the model selects the most popular meaning (if several), according to its commonness value. `Upper bound` is an oracle, which always predicts correct meaning if it is *nil* or is among top 20 most frequent mention meanings. GLOW-based methods select meanings from the same set, thus `upper bound` shows the best quality our approach may achieve.

---

[1] https://github.com/ispras-texterra/ainl-2018-d2w-dataset.
[2] Note, that token embedding size is 101 = 100+ extra position to encode END_TOKEN. Similar idea is for concept embedding size and START_CONCEPT/END_CONCEPT.

Table 2. Hyperparameters.

Section	Parameter	Label	Value
3.2	Window size around mentions	w	100
4.1	Token embedding size		101
	Concept embedding size		102
	Fully-connected layers F_c, F_h size		500
	Decoder LSTM size		500
	Fully-connected layer F_s size		102
	Attention		Bahdanau [1]
	Attention size		500
	Loss function		cosine distance
	Optimizer		Nadam
	Batch size		16
4.2	Forward/backward LSTM size	l	500
	Number of epochs		1110
4.3	Fully-connected layer F_l size		500
	Size of gated units stack	k	3
	CNN filter size		5
	Number of CNN filters		500
	Gated unit bias b size		500
	Number of epochs		2131
4.2 and 4.3	Keeping probability (train)	p	0.7

Implementation of GLOW system for Russian is fairly significant, as the results for this model outperform MCS by more than 4 percentage points. We additionally found out that GLOW without linker (which simply accepts top-scored concepts returned by ranker) performs even better. Applying features based on CNN and biLSTM generated smart context further improves the results.

Table 3. Evaluation results.

Model	Macro-averaged accuracy, %
MCS	83.01
Upper bound	94.03
GLOW	87.80
GLOW$_{no\_linker}$	88.01
GLOW + smart context features (CNN)	88.04
GLOW$_{no\_linker}$ + smart context features (CNN)	88.25
GLOW + smart context features (biLSTM)	88.19
GLOW$_{no\_linker}$ + smart context features (biLSTM)	**88.30**

To prove usefulness of the proposed features we split test data into 10 parts and evaluate baselines (GLOW and $GLOW_{no\_linker}$) and our best approach ($GLOW_{no\_linker}$ with biLSTM-based smart context features) on each part separately. Application of Wilcoxon signed rank test shows that our approach is better than baselines and the results are statistically significant with p-value <0.002.

6 Conclusion

In the current paper we propose a novel approach for generating context for the D2W task. Our method implies encoder-decoder architecture for translating sequence of tokens into concept embeddings.

During the research we trained two models with CNN and biLSTM based encoders and then compared their performance with the GLOW approach [18] implemented as baseline. Both of them outperform GLOW.

Despite moderate quality improvement of the result for the D2W task, we still consider the idea of translating tokens into concepts legible and expect to implement the encoder-decoder approach not only for D2W but for the whole Wikification task. Another line of work is to evaluate our approach on standard datasets for the English language.

References

1. Bahdanau, D., Cho, K., Bengio, Y.: Neural machine translation by jointly learning to align and translate. In: Proceedings of 3rd International Conference for Learning Representations, San Diego, pp. 1–15 (2015)
2. Bojanowski, P., Grave, E., Joulin, A., Mikolov, T.: Enriching word vectors with subword information. Trans. Assoc. Comput. Linguist. **5**, 135–146 (2017)
3. Cheng, X., Roth, D.: Relational inference for wikification. In: Proceedings of the 2013 Conference on Empirical Methods in Natural Language Processing. pp. 1787–1796 (2013)
4. Cho, K., Van Merriënboer, B., Bahdanau, D., Bengio, Y.: On the properties of neural machine translation: encoder-decoder approaches. In: Eighth Workshop on Syntax, Semantics and Structure in Statistical Translation (SSST-8) (2014)
5. Dandala, B., Mihalcea, R., Bunescu, R.: Word sense disambiguation using wikipedia. In: Gurevych, I., Kim, J. (eds.) The People's Web Meets NLP, pp. 241–262. Springer, Heidelberg (2013). https://doi.org/10.1007/978-3-642-35085-6_9
6. Durrett, G., Klein, D.: A joint model for entity analysis: coreference, typing, and linking. Trans. Assoc. Comput. Linguist. **2**, 477–490 (2014)
7. Francis-Landau, M., Durrett, G., Klein, D.: Capturing semantic similarity for entity linking with convolutional neural networks. In: Proceedings of NAACL-HLT, pp. 1256–1261 (2016)
8. Ganea, O.E., Hofmann, T.: Deep joint entity disambiguation with local neural attention (EMNLP 2017). In: Proceedings of the 2017 Conference on Empirical Methods in Natural Language Processing, pp. 2619–2629. Association for Computational Linguistics (2017)

9. Gehring, J., Auli, M., Grangier, D., Dauphin, Y.: A convolutional encoder model for neural machine translation. In: Proceedings of the 55th Annual Meeting of the Association for Computational Linguistics: Long Papers, vol. 1, pp. 123–135 (2017)

10. Gehring, J., Auli, M., Grangier, D., Yarats, D., Dauphin, Y.N.: Convolutional sequence to sequence learning. In: International Conference on Machine Learning, pp. 1243–1252 (2017)

11. Guo, Z., Barbosa, D.: Robust named entity disambiguation with random walks. Semant. Web 1–21 (2016, preprint)

12. Joachims, T.: Optimizing search engines using clickthrough data. In: Proceedings of the Eighth ACM SIGKDD International Conference on Knowledge Discovery and Data Mining, pp. 133–142. ACM (2002)

13. Li, J., Cai, Y., Cai, Z., Leung, H., Yang, K.: Wikipedia based short text classification method. In: Bao, Z., Trajcevski, G., Chang, L., Hua, W. (eds.) DASFAA 2017. LNCS, vol. 10179, pp. 275–286. Springer, Cham (2017). https://doi.org/10.1007/978-3-319-55705-2_22

14. Luong, T., Pham, H., Manning, C.D.: Effective approaches to attention-based neural machine translation. In: Proceedings of the 2015 Conference on Empirical Methods in Natural Language Processing, pp. 1412–1421 (2015)

15. Mikolov, T., Sutskever, I., Chen, K., Corrado, G., Dean, J.: Distributed representations of words and phrases and their compositionality. In: Advances in Neural Information Processing Systems, pp. 3111–3119 (2013)

16. Milne, D., Witten, I.H.: Learning to link with Wikipedia. In: Proceedings of the 17th ACM Conference on Information and Knowledge Management, pp. 509–518. ACM (2008)

17. Nothman, J., Ringland, N., Radford, W., Murphy, T., Curran, J.R.: Learning multilingual named entity recognition from Wikipedia. Artif. Intell. **194**, 151–175 (2013)

18. Ratinov, L., Roth, D., Downey, D., Anderson, M.: Local and global algorithms for disambiguation to Wikipedia. In: Proceedings of the 49th Annual Meeting of the Association for Computational Linguistics: Human Language Technologies, vol. 1, pp. 1375–1384. Association for Computational Linguistics (2011)

19. Sutskever, I., Vinyals, O., Le, Q.V.: Sequence to sequence learning with neural networks. In: Advances in Neural Information Processing Systems, pp. 3104–3112 (2014)

20. Sysoev, A., Andrianov, I.: Named entity recognition in Russian: the power of wiki-based approach. In: Proceedings of International Conference "Dialogue-2016", pp. 746–755 (2016)

21. Turdakov, D., et al.: Semantic analysis of texts using texterra system (2014). http://www.dialog-21.ru/digests/dialog2014/materials/pdf/TurdakovDY.pdf. Accessed 28 May 2018

22. Wu, G., He, Y., Hu, X.: Entity linking: an issue to extract corresponding entity with knowledge base. IEEE Access **6**, 6220–6231 (2018)

23. Yamada, I., Ito, T., Takeda, H., Takefuji, Y.: Linkify: enhancing text reading experience by detecting and linking helpful entities to users. IEEE Intell. Syst. (2018)

24. Zhou, J., Cao, Y., Wang, X., Li, P., Xu, W.: Deep recurrent models with fast-forward connections for neural machine translation. Trans. Assoc. Comput. Linguist. **4**(1), 371–383 (2016)

A Multi-feature Classifier for Verbal Metaphor Identification in Russian Texts

Yulia Badryzlova[1]([⊠]) [iD] and Polina Panicheva[2] [iD]

[1] National Research University Higher School of Economics, Moscow, Russia
yuliya.badryzlova@gmail.com
[2] St. Petersburg State University, Saint Petersburg, Russia
ppolin86@gmail.com

Abstract. The paper presents a supervised machine learning experiment with multiple features for identification of sentences containing verbal metaphors in raw Russian text. We introduce the custom-created training dataset, describe the feature engineering techniques, and discuss the results. The following set of features is applied: distributional semantic features, lexical and morphosyntactic co-occurrence frequencies, flag words, quotation marks, and sentence length. We combine these features into models of varying complexity; the results of the experiment demonstrate that fairly simple models based on lexical, morphosyntactic and semantic features are able to produce competitive results.

Keywords: Sentence-level metaphor identification
Supervised binary classification · Feature engineering
Distributional semantic features · Lexical co-occurrence features
Morphosyntactic co-occurrence features

1 Introduction

Metaphor is said to be a ubiquitous yet a fugitive phenomenon: it resides in virtually every utterance of human language, but it is notoriously difficult to formalize. Not only is metaphor indispensable in various language processing tasks; it is also commonly accepted that metaphor is a pervasive process in human language and thought [20], with numerous effects in psychology, psycholinguistics, and cognitive disciplines.

Metaphor processing has attracted increasing attention and effort in recent years. A series of Workshops on Metaphor in NLP was held for several successive years as a part of the NAACL-HLT conference. The most comprehensive overview of approaches to automated metaphor identification is available in [41].

The following types of features are exploited in the state-of-the-art systems for metaphor identification in the supervised and the unsupervised settings:

- lexical [3, 4, 10, 15, 17, 22, 23, 28, 30];
- morphological [4, 15];
- distributional semantic [28, 31, 35, 38];
- topic modelling [4, 13];

© Springer Nature Switzerland AG 2018
D. Ustalov et al. (Eds.): AINL 2018, CCIS 930, pp. 23–34, 2018.
https://doi.org/10.1007/978-3-030-01204-5_3

- lexical thesauri and ontologies: WordNet [3, 10, 15, 17, 25, 27, 28, 33, 38, 39], FrameNet [11], VerbNet [3], ConceptNet [30], and the SUMO ontology [8, 9];
- psycholinguistic features [3, 10, 28, 29, 37–40];
- syntactic relations [15, 30].

Metaphor identification projects can be divided into two groups according to their theoretical premises. Experiments in the first group stem from the conceptual metaphor paradigm [20] which stipulates that linguistic metaphors are surface realizations of the underlying conceptual mappings between the source and the target domains. Projects of this type seek to identify evidence of such mappings in the text [e.g. 8–11, 13, 25, 27, 28, 30, 37]. The second vein of experimental research does not make any *a priori* assumptions about the underlying conceptual mechanisms of metaphor and searches for any stretches of metaphoric language in the text [e.g. 3, 4, 15, 17, 29, 31, 33, 35, 38–40].

Results of metaphor identification experiments are difficult to compare for a number of reasons: (a) the theoretical incompatibility and the subsequent differences in the experimental design; (b) some systems identify metaphors on the sentence level while others identify word-level metaphors; (c) many of the existing systems are domain-specific; and (d) most systems are trained and evaluated on different datasets.

Metaphor identification in Russian texts has been addressed in several projects. For example, [28, 30, 37] use a variety of features to model the conceptual source and target domains and to align them with their linguistic realizations in text, while [31, 38, 39] operate outside of the conceptual metaphor paradigm. The former two systems exploit cross-linguistic metaphors: the classifier is first trained on the English data, and then the trained model is projected to Russian using a dictionary. The latter project uses distributional semantic vectors to distinguish metaphoric and non-metaphoric sentences.

The subsequent sections of this paper describe a sentence-level Russian verbal metaphor identification experiment on raw text with a rich multi-feature classifier involving semantic, lexical, and morphological features, as well as information about the occurrence of flag words (specific lexical markers), quotation marks, and sentence length.

To the best of our knowledge, this is the first project outside of the conceptual metaphor paradigm to explore a model of such complexity for metaphor identification in Russian texts.

2   The Dataset

The experimental dataset is comprised of 7,166 sentences each of which contains one of the 20 polysemous Russian verbs (referred to as target verbs below); some of the experimental verbs are listed in Table 1. The full dataset and its description are available for download.[1]

---

[1] https://github.com/yubadryzlova/metaphor_dataset_20_verbs.git.

Table 1. Dataset: some of the target verbs

Russian	Transliteration	Translation (primary meaning)
бомбардировать	bombardirovat	to bombard (smth/smb))
доить	doit	to milk (e.g. a cow)
нападать	napadat	to attack (smth/smb)
отрубить	otrubit	to hack (smth) off
трубить	trubit	to blow a trumpet
уколоть	ukolot	to prick (smth/smb)
зажигать	zazhigat	to ignite (smth)

2.1 The Target Verbs

The verbs were chosen so as to match the specific linguistic properties:

- the verb has at least one primary meaning which is a typical meaning of Accomplishment or Activity [26, 42];
- the verb has at least one primary meaning which authorises a two-actant construction with the following mandatory actants: (1) the Agent, (2) the Patient/the Theme;
- the Agent denotes a human being(s); the other actants refer to physical (concrete, non-abstract) entities;
- the derivational structure of the verb's polysemy is transparent: each secondary meaning is derived from the primary one by means of either a metaphoric or a distant metonymic shift;
- the verb has a small number (<10) of meanings listed in the dictionary;
- the verb does not possess any strongly delexicalized meanings.

Verbs of this kind were chosen for the experiment because they bring the opposition of metaphoric and non-metaphoric meanings to its most distinct expression.

2.2 The Non-metaphoric and the Metaphoric Classes

The sentences in the dataset are divided into the two classes, the non-metaphoric and the metaphoric ones.

The Non-metaphoric Class. This class includes the sentences where the target verb is used either (a) in the central literal meaning (as described above) or (b) in the meanings that are related to the central meaning via either a diathetic shift (i.e. the change of the syntactic rank of the actants), or a close metonymic shift.

The Metaphoric Class. This class contains the three types of sentences: (c) conventionalized metaphors based on polysemy, (d) unconventional creative metaphors, and (e) idiomatic expressions.

Conventionalized Metaphors are the target verbs used in their secondary meanings. For example, consider the metaphoric meanings of *trubit* 'to blow a trumpet'[2]:

- to talk profusely about smb, smth; to spread gossip, information, news, etc.;
- to perform a tiresome or tedious activity during a long period of time.

Unconventional Metaphors exploit the target verbs creatively to liken concepts from the target domain to concepts from the source domain [20] and to reinterpret the target in terms of the source, e.g. Сестра поглядела на нее, словно <уколола> кинжалом. 'Sister threw a glance at her, as if she <pricked> her with a dagger.'

Idiomatic Expressions are fixed or semi-fixed compositional units whose meaning is not equal to the sum of the meanings of its constituent lexemes, e.g. Когда то мои пра - пра - пра - пра - прадеды ... <грели руки> на ростовщичестве. 'There was a time when my fore- fore-fore-fore-fore-forefathers used to <warm their hands> (= to make dishonest or illegal profit) with usury.'

Sentence Selection and Annotation The sentences were obtained from RuTenTen11, a 14.5 bn-word Russian web corpus, accessed via the SketchEngine interface [16]. The sentences were added to the dataset in the order in which they were retrieved, without any filtering. The selection of sentences and their annotation by the binary classes (metaphoric vs. non-metaphoric) was performed by one annotator, a trained linguist. The annotator was compelled to make binary decisions.

The subsets for the individual verbs are balanced by the class, i.e. 50% of the sentences are metaphoric while the other half are non-metaphoric. However, the dataset is not balanced across the verbs (ranging between 225 and 693 sentences per verb). The data is heterogeneous in terms of genre and domain, containing non-normative Russian usage, which increases the difficulty of the classification task.

3 The Feature Set

3.1 Dataset Preprocessing and the Context Windows

The window-dependent features described below (the semantic, the lexical, and the morphosyntactic ones) were computed (a) on the fixed context windows of the sizes 2, 3, 4, and 5; (b) on the unfixed-size window equivalent to the length of the full sentence; and (c) on the set of the syntactic arguments of the target verb (its direct dependencies and some of their secondary projections).

Only content non-stopwords were included into the semantic and the lexical windows; as for the morphosyntactic windows, they were comprised of all the grammemes found within a given window, including prepositions and punctuation marks.

The syntactic arguments of the target verbs and the morphological characteristics of lemmas were obtained with the online interface for the Russian MaltParser [7].

---

[2] The definitions throughout the paper are quoted from the Dictionary of the Russian Language [44].

3.2 Distributional Semantic Features

The Word-Embeddings Models. Our semantic features are based on word-embeddings models. We experiment with two pre-trained models presented in [19] that are freely available for download from the RusVectōrēs website [34]; both models were trained with the word2vec Continuous Skipgram algorithm.

- The WikiRNC model was trained with vector dimensionality 300 and window size 2 on the joint corpus of Russian Wikipedia and the Russian National Corpus, with the total of 600 m tokens;
- The Araneum model was trained on a much larger corpus, Araneum Russicum Maximum [5], of about 10bn tokens, with vector dimensionality 600 and the window size of 2.

The Semantic Similarity Measure. When we apply distributional semantics to context windows of different sizes, we proceed from the intuition that a metaphoric verb will be semantically deviant from its linear context window, affecting the mean semantic similarity between the words in the window in a negative way, whereas a literally used verb will belong to the same conceptual domain as its context words, making the contextual sub-space denser and adding to the mean similarity [14].

Application of distributional semantic models to the syntactic arguments of the verbs relies on the consideration that metaphor is a Selectional Preference violation [43], which is effectively captured as semantic deviance between the metaphoric verb and its main arguments [35]. The assumption is that a verb used in a literal sense will belong to the same conceptual domain as its immediate arguments, whereas metaphoric verb usage implies arguments belonging to a different conceptual domain.

The semantic similarity of tokens within the context is calculated as the following:

$$Sim_{win} = Mean\{ \; Sim(w_i, w_j) | w_i, w_j \in Win \}, \tag{1}$$

$$SimV_{win} = Mean\{ \; Sim(w_i; w_j) | w_i, w_j \in Win, w_i \neq verb, w_j \neq verb \}, \tag{2}$$

$$SimDiff_{win} = Sim_{win} - SimV_{win,} \tag{3}$$

where *Sim* is the semantic similarity in the distributional semantic space, and *Win* is the context window around the target verb: a linear window in the case of linear context, or the list of syntactic arguments in the case of the syntactic arguments context.

The Augmented Semantic Features. If a sentence in our corpus features a low-frequency word that is missing from the model, its measure of semantic similarity with its environment equals to zero. We moderated this effect by replacing the unavailable similarities by the mean of all the similarity measures in the current context window.

3.3 Lexical Co-occurrence Features

The use of lexical features for metaphor identification draws on the notion of lexico-semantic combinability [2], i.e. that different meanings of polysemous words impose restrictions on the semantics of their arguments, and subsequently, on their lexical classes. For example, the non-metaphoric meaning of *raspylyat* 'to spray' will often co-occur with lexemes from the class of liquids and powder-like substances (water, perfume, chemicals, and the like), while the metaphoric meaning 'to scatter, to disperse smth thus decreasing its efficiency' will typically co-occur with words denoting valuable resources (money, funds, effort, energy, troops, reserves, etc.).

To vectorize the unigrams of lemmas, we applied several measures of association: weirdness [1], the extension of Student's t-test proposed in [24], log likelihood [6], and Kullback-Leibler Divergence [18]. The best results were produced by the ΔP metric [21] which is calculated according to the formula:

$$\frac{a}{a + \neg a} - \frac{b}{b + \neg b}, \tag{4}$$

where a is the number of occurrences of a lexeme in the metaphoric subcorpus, b is the number of occurrences of the lexeme in the non-metaphoric corpus, $a + \neg a$ is the size of the metaphoric subcorpus, and $b + \neg b$ is the size of the non-metaphoric subcorpus.

3.4 Morphosyntactic Co-occurrence Features

The rationale behind the use of morphosyntax in metaphor identification is grounded in the fact that different meanings of a polysemous verb may develop exclusive morphosyntactic constructions. For example, in the verb *otrubit* (whose non-metaphoric meaning is 'to hack smth off'), the metaphoric meaning ('to respond, to say smth in a brusque or abrupt manner') develops an intransitive construction; this meaning is often used to introduce direct speech in the narration:

— *Нет,* — <*отрубил*> *Керк.* — *Деньги должны быть выиграны сегодня.* 'No', Kirk <cut off> (= responded abruptly), 'the money must be won today'.

We explored three different configurations of morphological characteristics of nouns and verbs which vary in the fullness of representation:

1. verb only pos/noun only pos: indication of only the part of speech;
2. verb full: part of speech, aspect, tense, number, mood, gender, and person;
3. noun full: part of speech, gender, animacy, case, and number;
4. verb short: part of speech, aspect, tense, mood;
5. noun short: part of speech, animacy, case;

We tested five combinations of morphological configurations: verb only pos + noun only pos, verb full + noun full, verb short + noun full, verb full + noun short, and verb short + noun short. Prepositions and punctuation in all the configurations were represented by their lemmas; all the other parts of speech were always represented by their POS tags.

Besides, we experimented with unigrams, bigrams, and trigrams of morphosyntactic tags: bigrams and trigrams are expected to capture the linear order of grammemes in the context window, while unigrams show their distribution in sentences irrespective of the linear order. The association measure between grammemes on the one hand, and the non-metaphoric/metaphoric class on the other was calculated with the ΔP metric.

4   Experimental Setup

The metaphor identification task was formulated as sentence-level binary classification: the classifier was to identify which sentences belonged to the metaphoric and the non-metaphoric classes. We experimented with the datasets of individual verbs and with the combined dataset of all the 20 verbs.

We used the Support Vector Machine (SVM) classifier with linear kernel[3]; the experiments were run using 5-fold cross-validation.

We experimented with a total of 45 models, i.e. with different one-, two-, three-, four-, and five-feature combinations.

The results of the performance were estimated as the accuracy of classification.

5   Results

5.1   Features' Impact

Beside the features described in Sects. 3.2–3.4, we also tested the following features: (a) specific lexical markers of metaphoricity ('flag words', see [12, 36]); (b) quotation marks; and (c) sentence length. However, none of them proved efficient, either in isolation or in combination with the other features.

All the window-dependent features (semantic, lexical, and morphosyntactic) have proved to be quite sensitive to the size of the context window. Figure 1 demonstrates the correlation between the classification results (accuracy) on the lexical features and the size of the window for three the verbs which demonstrate a downward, an upward, and a flat dynamics.

Obviously, this behaviour is connected with the distances at which the lemmas with conspicuous association scores occur in relation to the target verb.

For example, *otrubit* 'to hack smth off' best performs on the set of the syntactic arguments; this is due to the high frequency of the metaphoric intransitive construction which serves to introduce direct speech (see Sect. 3.4). The verb in this construction has only one syntactic argument, the subject, which is typically a person's name. Proper names are low-frequency lemmas, and therefore they will have low association scores. Whereas the non-metaphoric meaning tends to co-occur with higher-frequency syntactic arguments on a much more regular basis (e.g. 'to cut off a chunk of wood/smb's head', etc.); these lemmas have high association scores. This contrast imparts high predictive power to the model based on the syntactic arguments of *otrubit*. Using linear windows, especially larger ones, introduces excessive noise into the model and disorients the classifier.

However, in the aggregate terms across the dataset, the large-size windows (full_sent and win5) by far outperform the other windows.

The accuracies of the non-augmented models and their augmented counterparts showed no significant difference.

---

[3] LinearSVC, as implemented in scikit-learn [32].

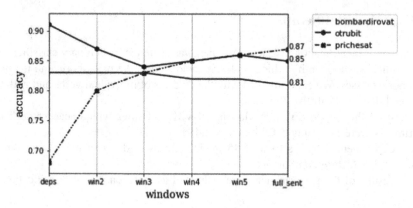

Fig. 1. Correlation between the accuracy of classification and the size/type of the context window (lexical co-occurrence features). '*Deps*' – the set of the verb's syntactic arguments; '*win2*' – '*win5*' – windows of the sizes 2–5; '*full_sent*' – window of the full sentence length.

The morphologically poor configuration of grammemes ('only pos') is demonstrably outperformed by the morphologically informed configurations (2–5, see the list in Sect. 3.4). Meanwhile, there is no pronounced leader among the morphologically informed configurations: they all perform at approximately the same level.

Besides, morphosyntactic unigrams consistently outperform trigrams, while being almost on a par with bigrams.

In sum, the efficient models provided by our features are the one-, two-, and three-feature combinations of the semantic, the lexical, and the morphosyntactic features.

5.2   Classification Results

We report the results for the models with the following options:

- the features are computed on the full sentence window;
- the distributional semantic feature ('sem') is the non-augmented version computed on the Araneum word-embeddings model;
- the morphosyntactic feature ('morph') is computed on unigrams of the configuration 'verb full + noun full';
- the lexical co-occurrence feature ('lex') is computed as described in Sect. 3.3.

An abridged version of the classification results is presented in Table 2. The full version of the table can be accessed online (See footnote 1).

The accuracies of the models across the verbs range within the following limits: 'sem': 0.52–0.81; 'lex': 0.77–0.94; 'morph': 0.67–0.82; 'sem+lex': 0.77–0.94; 'sem+morph': 0.7–0.85; 'lex+morph': 0.77–0.96; 'sem+lex+morph': 0.75–0.95.

The best accuracies on individual verbs range from the moderate 0.77 to the quite encouraging 0.96. The accuracy of the classifier on the combined dataset of the 20 verbs reached the mark of 0.83. This performance is on a competitive footing with the results reported by the other systems for metaphor identification in Russian: the F-scores of 0.76 in [39] and 0.84 in [38] which use the translation method and experiment

Table 2. Accuracy of classification (selected verbs)

dataset / model	sem	lex	morph	sem+lex	sem+morph	lex+morph	sem+lex+ morph
bombardirovat	0.75	0.81	0.75	0.83	0.77	0.82	<u>0.85</u>
napadat	0.59	<u>0.77</u>	0.76	0.77	0.74	0.77	0.75
vykraivat	0.81	0.94	0.82	0.93	0.85	<u>0.96</u>	0.95
combined dataset (20 verbs)	0.65	0.82	0.67	0.82	0.71	<u>0.83</u>	0.83

with much smaller datasets of pre-filtered SVO triples and adjective-noun tuples; and the accuracy of 0.68 in [31] which is run in a setting comparable to ours.

In five of the 20 verbs, the best result is achieved with the simple model 'lex'; adding further features does not lead to a gain in efficiency. The composite model of the semantic and the lexical features ('sem+lex') yields the best result only in two verbs. The majority of the top accuracies is achieved with the combination of the two features, the lexical and the morphosyntactic ones, ('lex+morph') – in 10 of the individual verbs, and on the joint dataset. In four individual verbs, the best results are obtained with the most complex model composed of the three features, the semantic, the lexical, and the morphosyntactic ones ('sem+lex+morph'). On the joint dataset, the last two models yield an identical result.

Interestingly, the 'sem', the 'morph', and the 'sem+morph' models consistently fall behind the other models across the datasets, as morphology alone cannot be expected to reliably predict the metaphoric or the non-metaphoric class. As for the comparatively low efficiency of the distributional semantic feature, it presumably can be accounted for by the fact that state-of-the-art distributional semantic models do not discriminate between different meanings of polysemous words; they generate a single vector which collapses all the senses of a word into a single value. The classification results will depend on the nature of the typical senses of the target verb and their co-occurrences in the training corpus (a fact also addressed in [31]).

To summarize, we can say that on a dataset composed of multiple target verbs, the two models are most likely to produce the high accuracy result: the two-feature combination 'lex+morph', and the three-feature combination 'sem+lex+morph'.

However, this observation may hold true only for verbs that are characterised by the semantic and the actant structure properties described in Sect. 2.1.

6 Conclusion

We have presented a manually annotated experimental dataset of metaphoric and non-metaphoric sentences featuring 20 target verbs. We also introduced the set of experimental features and presented their linguistic motivation. Next, we described the setup of the experiment for classifying the sentences into the metaphoric and the non-metaphoric classes. The results of the experiment suggest that the two composite models are likely to be scalable: the model combining the lexical and the morphosyntactic features, and the model based on the combination of the semantic, lexical, and morphosyntactic features. However, this generalization may hold true only for verbs of the same type as the target verbs in the experimental dataset (i.e. typical Activity or Accomplishment verbs with two actants).

Acknowledgements. The contribution to this study by Polina Panicheva is supported by RFBR grant № 16-06-00529.

References

1. Ahmad, K., Gillam, L., Tostevin, L.: Weirdness indexing for logical document extrapolation and retrieval (WILDER). In: Voorhees, E., Harman, D. (eds.) Proceedings of the 8th Text Retrieval Conference, TREC 8, Gaithersburg, MA, pp. 717–724 (2000)
2. Apresjan, Yu. D.: Izbrannyye trudy, t.1. Leksicheskaya semantika. Sinonimicheskiye sredstva yazyka/Selected works, vol. 1. Lexical semantics. The synonymic means of the language, 2nd edn. LRC Publishing House, Moscow (1995)
3. Beigman Klebanov, B., Leong, C.W., Gutierrez, E.D., Shutova, E., Flor, M.: Semantic classifications for detection of verb metaphors. In: Proceedings of the 54th Annual Meeting of ACL 2016, Berlin, Germany, vol. 2, pp. 101–106 (2016)
4. Beigman Klebanov, B., Leong, C.W., Heilman, M., Flor, M.: Different texts, same metaphors: unigrams and beyond. In: Proceedings of the Second Workshop on Metaphor in NLP, Baltimore, MD, pp. 11–17 (2014)
5. Benko, V., Zakharov, V.: Very large Russian corpora: new opportunities and new challenges. In: Selegey, V. (ed.) Proceedings of the Annual International Conference "Dialogue", Moscow, Russia, pp. 83–98 (2016)
6. Cressie, N., Read, T.R.C.: Multinomial goodness-of-fit tests. J. R. Stat. Soc. Ser. B (Methodol.) **46**(3), 440–464 (1984)
7. Droganova, K.A., Medyankin, N.S.: NLP pipeline for Russian: an easy-to-use web application for morphological and syntactic annotation. In: Proceedings of the Annual International Conference "Dialogue", Moscow, Russia (2016)
8. Dunn, J.: Evaluating the premises and results of four metaphor identification systems. In: Proceedings of CICLing 2013, Samos, Greece, pp. 471–486 (2013a)
9. Dunn, J.: What metaphor identification systems can tell us about metaphor-in-language. In: Proceedings of the First Workshop on Metaphor in NLP, Atlanta, GA, pp. 1–10 (2013b)
10. Gandy, L., et al.: Automatic identification of conceptual metaphors with limited knowledge. In: Proceedings of the 27th AAAI Conference on Artificial Intelligence, pp. 328–334. AAAI Press, Bellevue, WA (2013)

11. Gedigian, M., Bryant, J., Narayanan, S., Ciric, B.: Catching metaphors. In: Proceedings of the Third Workshop on Scalable Natural Language Understanding, ScaNaLU 2006, New York City, pp. 41–48 (2006)
12. Goatly, A.: The Language of Metaphors. Routledge, Abingdon (2011)
13. Heintz, I., et al.: Automatic extraction of linguistic metaphors with LDA topic modeling. In: Proceedings of the First Workshop on Metaphor in NLP, Atlanta, GA, pp. 58–66 (2013)
14. Herbelot, A., Kochmar, E.: 'Calling on the classical phone': a distributional model of adjective-noun errors in learners' English. In: Proceedings of COLING 2016, Osaka, Japan, pp. 976–986 (2016)
15. Hovy, D., et al.: Identifying metaphorical word use with tree kernels. In: Proceedings of the First Workshop on Metaphor in NLP, Atlanta, GA, pp. 52–57 (2013)
16. Kilgarriff, A., et al.: The Sketch Engine: ten years on. Lexicography 1, 7–36 (2014)
17. Krishnakumaran, S., Zhu, X.: Hunting elusive metaphors using lexical resources. In: Proceedings of the Workshop on Computational Approaches to Figurative Language, NAACL-HLT 2007, Rochester, NY, pp. 13–20 (2007)
18. Kullback, S., Leibler, R.A.: On Information and Sufficiency. Annals of Mathematical Statistics, vol. 22(1), pp. 79–86. Institute of Mathematical Statistics, Ann Arbor (1951)
19. Kutuzov, A., Kuzmenko, E.: WebVectors: a toolkit for building web interfaces for vector semantic models. In: Ignatov, D.I., et al. (eds.) AIST 2016. CCIS, vol. 661, pp. 155–161. Springer, Cham (2017). https://doi.org/10.1007/978-3-319-52920-2_15
20. Lakoff, G., Johnson, M.: Metaphors We Live By. The University of Chicago Press, Chicago (1980)
21. Levshina, N.: How to Do Linguistics with R: Data Exploration and Statistical Analysis. John Benjamins Publishing Company, Amsterdam (2015)
22. Li, L., Sporleder, C.: Classifier combination for contextual idiom detection without labelled data. In: Proceedings of EMNLP 2009, Singapore, pp. 315–323 (2009)
23. Li, L., Sporleder, C.: Using gaussian mixture models to detect figurative language in context. In: Proceedings of NAACL HLT 2010, Los Angeles, CA, pp. 297–300 (2010)
24. Manning, C.D., Schütze, H.: Foundations of Statistical Natural Language Processing. MIT Press, Cambridge (1999)
25. Mason, Z.J.: CorMet: a computational, corpus-based conventional metaphor extraction system. Comput. Linguist. 30(1), 23–44 (2004)
26. Mehlig, H.R.: Semantika predlozheniya i semantika vida v russkom yazyke/The semantics of the sentence and the semantics of the aspect in the Russian language. In: Bulygina, T.V., Kibrik, A.E. (eds.) Novoye v zarubezhnoy lingvistike, vyp. 15. Sovremennaya zarubezhnaya lingvistika/The state-of-the-art in international linguistics, issue 15. Contemporary international linguistics, pp. 227–249. Progress, Moscow (1985)
27. Mohler, M., Bracewell, D., Tomlinson, M., Hinote, D.: Semantic signatures for example-based linguistic metaphor detection. In: Proceedings of the First Workshop on Metaphor in NLP, Atlanta, GA, pp. 27–35 (2013)
28. Mohler, M., Rink, B., Bracewell, D.B., Tomlinson, M.T.: A novel distributional approach to multilingual conceptual metaphor recognition. In: Proceedings of COLING 2014: Technical Papers, Dublin, Ireland, pp. 1752–1763 (2014)
29. Neuman, Y., et al.: Metaphor identification in large texts corpora. PLoS ONE 8(4), e62343 (2013)
30. Ovchinnikova, E., Israel, R., Wertheim, S., Zaytsev, V., Montazeri, N., Hobbs, J.: Abductive inference for interpretation of metaphors. In: Proceedings of the 2nd Workshop on Metaphor in NLP, NAACL-HLT 2014, Denver, CO, pp. 33–41 (2014)

31. Panicheva, P., Badryzlova, Yu.: Distributional semantic features in Russian verbal metaphor identification. In: Selegey, V. (ed.) Proceedings of the Annual International Conference "Dialogue", vol. 1, pp. 179–190. Moscow, Russia (2017)
32. Pedregosa, F., et al.: Scikit-learn: machine learning in Python. J. Mach. Learn. Res. **12**, 2825–2830 (2011)
33. Peters, W., Peters, I.: Lexicalised systematic polysemy in WordNet. In: Proceedings of LREC 2000, Athens, Greece (2000)
34. RusVectōrēs: Word Embeddings for Russian Online. http://rusvectores.org/ru/models/. Accessed 01 July 2018
35. Shutova, E., Kiela, D., Maillard, J.: Black holes and white rabbits: metaphor identification with visual features. In: Proceedings of NAACL HLT 2016, pp. 160–170 (2016)
36. Steen, G.J., Dorst, L., Herrmann, B., Kaal, A., Krennmayr, T., Pasma, T.: A Method for Linguistic Metaphor Identification: from MIP to MIPVU. John Benjamins Publishing, Amsterdam (2010)
37. Strzalkowski, T., et al.: Robust extraction of metaphor from novel data. In: Proceedings of the First Workshop on Metaphor in NLP, Atlanta, GA, pp. 67–76 (2013)
38. Tsvetkov, Y., Boytsov, L., Gershman, A., Nyberg, E., Dyer, C.: Metaphor detection with cross-lingual model transfer. In: Proceedings of the 52nd Annual Meeting of ACL, vol. 1, Baltimore, MD, pp. 248–258 (2014)
39. Tsvetkov, Y., Mukomel, E., Gershman, A.: Cross-lingual metaphor detection using common semantic features. In: Proceedings of the First Workshop on Metaphor in NLP, Atlanta, GA, pp. 45–51 (2013)
40. Turney, P.D., Neuman, Y., Assaf, D., Cohen, Y.: Literal and metaphorical sense identification through concrete and abstract context. In: Proceedings of EMNLP 2011, Edinburgh, UK, pp. 680–690 (2011)
41. Veale, T., Shutova, E., Beigman Klebanov, B.: Metaphor: a computational perspective. In: Synthesis Lectures on Human Language Technologies, vol. 9, no. 1, pp. 1–160 (2016)
42. Vendler, Z.: Verbs and times. Philos. Rev. **66**, 143–160 (1957)
43. Wilks, Y.: Making preferences more active. Artif. Intell. **11**(3), 197–223 (1978)
44. Yevgenyeva, A.P. (ed.): Slovar russkogo yazyka v 4 tomakh/The dictionary of the Russian language in 4 volumes, 2nd edn. The Institute for the Russian Language, Moscow (1981–1984)

Lemmatization for Ancient Languages: Rules or Neural Networks?

Oksana Dereza[1,2]([⊠]) [iD]

[1] National Research University "Higher School of Economics", Moscow, Russia
odereza@hse.ru
[2] Lomonosov Moscow State University, Moscow, Russia
https://www.hse.ru/en/staff/odereza

Abstract. Lemmatisation, which is one of the most important stages of text preprocessing, consists in grouping the inflected forms of a word together so they can be analysed as a single item. This task is often considered solved for most modern languages irregardless of their morphological type, but the situation is dramatically different for ancient languages. Rich inflectional system and high level of orthographic variation common to these languages together with lack of resources make lemmatising historical data a challenging task. It becomes more and more important as manuscripts are being extensively digitized now, but still remains poorly covered in literature. In this work, I compare a rule-based and a neural network based approach to lemmatisation in case of Early Irish (Old and Middle Irish are often described together as "Early Irish") data.

Keywords: Early Irish · Natural language processing
Under-resourced languages · Lemmatisation · Neural networks
Sequence-to-sequence learning

1 Introduction

Lemmatisation, which is one of the most important stages of text preprocessing, consists in grouping the inflected forms of a word together so they can be analysed as a single item, identified by the word's lemma, or dictionary form. It is not a very complicated task for languages such as English, where a paradigm consists of a few forms close in spelling; but when it comes to morphologically rich languages, such as Russian, Hungarian or Irish, lemmatisation becomes more challenging. However, this task is often considered solved for most resource-rich modern languages irregardless of their morphological type. The situation is dramatically different for ancient languages characterised not only by a rich inflectional system, but also by a high level of orthographic variation. Lemmatisation for ancient languages is still poorly covered in literature, although this task becomes more and more important as manuscripts are being extensively digitized.

© Springer Nature Switzerland AG 2018
D. Ustalov et al. (Eds.): AINL 2018, CCIS 930, pp. 35–47, 2018.
https://doi.org/10.1007/978-3-030-01204-5_4

There are two suitable approaches to this task that I will describe and compare in this article in regard to Early Irish data: a rule-based approach and character-based neural network models.

2   Related Works

The problem of NLP for historical languages first arose in the last quarter of the XXth century in regard to Ancient Greek [32], Sanskrit [20,47] and Latin [29,33] and for a long time was confined to these languages. As more and more medieval manuscripts were being digitised, there appeared a number of works dedicated to spelling variation in historical corpora, its normalisation and further linguistic processing for Early Modern English [3,4], Old French [44], Old Swedish [6], Early New High German [5], historical Portuguese [17,19,39], historical Slovene [40], Middle Welsh [30] and Middle Dutch [24,25]. Historical data processing in general has been surveyed in a substantial monograph [37] and several articles [16,36]. Apart from corpus studies, there have emerged several open-source tools for historical language processing, such as a Classical Language Toolkit[1] [22], which offers NLP support for the languages of Ancient, Classical, and Medieval Eurasia. For the moment, only Greek and Latin functionality in CLTK includes lemmatisation.

Lemmatisation has also been an active area of research in computational linguistics, especially for morphologically rich languages [8,9,12,13,18,28,43,46].

There are two major approaches to lemmatisation, a rule-based approach and a statistical one. The rule-based approach, which requires much manual intervention but yield very good results due to being language-specific, is widely used, examples being Swedish [11], Icelandic [21], Czech [23], Slovene [38], German [35], Hindi [34], Arabic [1,15] and many other languages. A classical work on automatic morphological analysis of Ancient Greek describes a stem lexicon, where each stem is marked with inflectional class, and a list of pseudo-suffixes needed to restore these stems to lemmas [32]. A Latin lemmatiser from the aforementioned Python library CLTK also uses stem and suffix lexicons. The best morphological analyser for Russian, Mystem, is based on Zalizniak grammatical dictionary [50]. This dictionary contains a detailed description of ca. 100,000 words that includes their inflectional classes. Mystem analyses unknown words by comparing them to the closest words in its lexicon. The 'closeness' is computed using the built-in suffix list [42]. A morphological analyser of modern Irish used in New Corpus of Ireland is based on finite-state transducers and described in [14] and [26].

Statistical approach to lemmatisation is computationally expensive and requires a large annotated corpus to train a model, especially when one deals with a complex inflectional system. Nevertheless, there are a few statistical parsers that achieve excellent results. Morfette, which was developed specially for fusional and agglutinative languages, simultaneously learns lemmas and PoS-tags using maximum entropy classifiers. It does not need hard-coded lists of

---

[1]   http://docs.cltk.org/en/latest/.

stems and suffixes and derives lemma classes itself from the working corpus [10]. It shows over 97% lemmatisation accuracy for seen words and over 75% accuracy for unseen words on Romanian, Spanish and Polish data. Another joint lemmatisation and PoS-tagging system, Lemming, achieves more than 93–98% for both known and unknown words on Czech, German, Spanish and Hungaian datasets [31]. Now there are models available for more than 15 languages, including Basque, Hebrew, Korean, Estonian, French and Arabic[2]. Unfortunately, it is almost impossible to directly compare the performance of rule-based and statistical-based systems for the same language described in different works due to the discrepancy of training datasets and the absence of evaluation results for some of the models.

Recently, neural networks also started being used for lemmatisation. For example, a system combining convolutional architecture that models orthography with distributional word embeddings that represent lexical context was successfully implemented by [25] to lemmatise Middle Dutch data. The authors obtained 94–97% accuracy for known words and 45–59% accuracy for unknown words on four different datasets.

3   Data

3.1   Sources

One of the most difficult problems one faces working on NLP tools for ancient languages is the lack of data. The quality of a machine learning model is widely known to depend upon the size of the training corpus. The only publicly available annotated corpus of Early Irish is POMIC [27], but it is not a very suitable source of data for machine learning because it is represented as parse trees in PSD format. Another substantial resource is the electronic edition of the Dictionary of the Irish Language[3] [45]. The DIL is a historical dictionary of Irish, which covers Old and Middle Irish periods. Each of 43,345 entries consists of a headword (lemma), a list of forms including different spellings and compounds and examples of use with a reference to source text.

However, the list of forms cited in the DIL is incomplete; apart from that, some of the forms are contracted: for example, the list of forms for *cruimther* 'priest' is represented in the dictionary as -ir, which the reader is to read as *cruimthir*, and the list of forms for *carpat* 'chariot' looks like *cairpthiu, -thib, -tiu, -tib* which has to be read as *cairpthiu, caipthib, cairptiu, cairptib*. Words can be abbreviated in many different ways, which is a consequence of the fact that there were many scholars who contributed to the DIL throughout 1913–1976, and each of them used his own notation, as preserved in the digital edition. Some common types of contractions are listed in Table 1.

Still, the DIL is the best source of data for training a lemmatiser. To compile a lexicon for the rule-based lemmatiser and a training corpus for the neural network

---

[2] http://cistern.cis.lmu.de/marmot/models/CURRENT/.
[3] http://dil.ie.

Table 1. Contracted, restored and missing forms and spellings from the DIL

DIL	Restored	Missing
carpat, cairpthiu, -thib, -tiu, -tib	carpat, cairpthiu, caipthib, cairptiu, cairptib	carbad, carbat, carbait, carpait, carput, carpti...
carat(r)as	caratas, caratras	caratrad, caradras, caradrus, caradruis, caratrais...
cruimther, -ir	cruimther, cruimthir	cruimter, crumther, cruimthear, crumper, crumpir, cromthar, crumthirech
anmothaig[thig]e	anmothaige, anmothige	anmothaigthech, anmotuighe...
aball, a.	aball	abhull, aboll, ubull, abaill, abla, abhla, ubla, ubhaill...

lemmatiser, I crawled DIL's website, parsed HTML files and derived a set of rules to restore contractions and remove unnecessary markup. As a result, I got 83,155 unique form-lemma pairs. They were then shuffled and split into training, validation and test sets, the former two being 5,000 samples each. One has to bear in mind, that this amount of training data is insufficient for getting extremely good results in lemmatisation for a language as morphologically complex and orthographically inconsistent as Early Irish.

Also, a test set was manually created to evaluate a rule-based system, because the DIL data cannot be used for evaluation in this case. It is described in detail in the next section.

3.2 Morphology and Orthography

Old Irish is a fusional language with an elaborate system of verbal and nominal inflexion, comparable to Ancient Greek and Sanskrit in its complexity. In Celtic languages, there are two ways to encode morphological information in a word form, which often occur together: regular endings and grammaticalised phonetic changes in the beginning of the word called 'initial mutations'. It means that the first sound of a word can change under specific grammatical conditions, for example, the word *céile* 'servant' with a definite article in nominative plural will take a form *ind chéili* 'the servants', where the first stop [k] mutated into fricative [x]. This type of mutation is called lenition, and in this particular case it shows the presence of a definite article in nominative plural masculine, while the ending *-i* means that the noun itself is in nominative plural. There are four types of initial mutations in Early Irish: lenition, eclipsis, t-prothesis and h-prothesis. I will not expand on how exactly they affect consonants and vowels and when they occur, because it is not relevant for the task. I have to mention though, that both in Old and Middle Irish mutations were inconsistently marked in writing, and the orthography on the whole involves much variation. There are several other orthographic features that increase a number of possible forms for a single lemma:

- inconsistent use of length marks;
- in later texts mute vowels indicate the neighbouring consonant's quality;
- complex verb forms can be spelled either with or without a hyphen or a whitespace.

Moreover, in Old and Middle Irish objective pronouns and relative particles are incorporated into a verb between the preverb and the root: cf. *caraid* 'he/she/it loves' and *rob-car-si* 'she has loved you', where *ro-* is a perfective particle, *-b-* is an infixed pronoun for 2nd person plural object, and *-si* is an emphatic suffixed pronoun 3rd person singular feminine. The presence of a preverb with dependent forms triggers a shift in stress, which causes complex morphophonological changes and often produces a number of very differently looking forms in a verbal paradigm, particularly in the case of compound verbs, cf. *do-beir* 'gives,brings' and *ní thabair* 'does not give, bring'. Table 2 illustrates the variety of Early Irish verbal forms through the example of *do-beir*.

Table 2. Some forms of the verb 'do-beir'

Form	Deuterotonic	Prototonic (after preverb)	Translation
INDIC PRES 3SG	do-beir	(ní) thabair	'does (not) give/bring'
SUBJ PRES 3SG	do-bera	(ní) thaibrea	'if does (not) give/bring'
PRET 3SG	do-bert	(ní) thubart	'did (not) give/bring'
FUT 3SG	do-béra	(ní) thibéra	'will (not) give/bring'
PERF 3SG	do-rat	(ní) tharat	'did (not) give'
PERF2 3SG	do-uic	(ní) thuicc	'did (not) bring'

I should also mention, that the DIL is not strictly grammatical in the following assumptions, and so are the models trained on it:

- verbal forms with infixed pronouns are lemmatised as verbal forms without a pronoun (*notbéra* 'will bring you' > *beirid* 'brings');
- compound forms of a preposition and a definite article are lemmatised as prepositions without an article (*isin* 'in + DET' > *i* 'in');
- prepositional pronouns are lemmatised as prepositions (*indtib* 'in them' > *i* 'in');
- emphatic suffixed pronouns (*-som, -siu, -si, -sa* etc.) are lemmatised as independent personal pronouns.

4   Rule-Based Approach

At first, I chose rule-based approach to lemma prediction over machine learning due to the scarcity of available data.

Morphophonological complexity of Early Irish compounded by the many non-transparent orthographic features makes traditional rule-based approach to lemmatisation with hard-coded lists of possible pseudo-suffixes and rules of their treatment less suitable for Early Irish than for other languages. A more reliable way for a start is building a full form lexicon where every word form corresponds to a lemma. I used the DIL described in the previous section for this purpose.

There was a series of experiments conducted on Early Irish prose texts that resulted into the following architecture of the rule-based lemmatiser. Every word in a text fed to the system is first demutated (i.e. the changes at the beginning of the word are eliminated) and then looked up in the dictionary. The lemmatiser returns a lemma for each known word and a demutated form for each unknown word by default; there is also an option to predict lemmas for unknown words with the help of Damerau-Levenshtein edit distance. For every unknown word, the program generates all possible strings on edit distance 1 and 2, checks them up in the dictionary and adds those that prove to be real words to the candidate list. Then the candidates are filtered by the first character: if the unknown word starts with a vowel, the candidate should also start with a vowel, and if the unknown word starts with a consonant, the candidate should start with the same consonant. Those parameters were chosen empirically as they yield the best results, i.e. the highest percentage of correctly predicted lemmas. Finally, the lemma of the candidate that has the highest probability is taken as a lemma for the unknown word.

In this work, I did not focus on word sense disambiguation, which means that if two or more different lemmas have identical forms, we cannot say for sure which lemma should be chosen for a particular instance of a homonymous form. The system provides two options for such cases: either return a list of all possible lemmas or choose the lemma with the highest probability. I should point out, that probability here is not a probability in a strict mathematical sense. Word form probability is formulated as a frequency count computed for each word in the test corpus, and lemma probability is the the sum of probabilities of forms belonging to a lemma.

Rule based lemmatiser was evaluated using accuracy score, which is a common metric for this task, on a set of manually annotated sentences, randomly chosen from Early Irish texts given in Table 3, which belong to different periods. The test set consists of 50 sentences, 840 tokens in total. It is worth mentioning, that the lemmatiser's lexicon contains mostly Old Irish forms with a small amount of Middle Irish ones and barely any Early Modern Irish ones. While Old Irish data helps to check how the system copes with unknown words' grammar, Middle and Early Modern Irish data is supposed to show its achievements with spelling variation. One also has to bear in mind, that the lemmatisers's performance is affected by form homonymy, which, given the absence of disambiguation, worsens the results.

Table 3. Early Irish texts used for creating a test set

Text	Period
Togail Bruidne Dá Derga	VII-IX centuries
Tochmarc Étaine	VIII-IX centuries
Fled Dúin na nGéd	XI-XII centuries
Lebor Gabála Érenn	XII century
Cath Finntrágha	XV century
Aided Muirchertaig Meic Erca	XIV-XV centuries
Buile Shuibhne	XVII-XVIII centuries

The system's performance is given in tables below; Table 4 compares the rule-based lemmatiser results with the baseline, defined as demutating a form, and Table 5 gives more detailed information.

Table 4. Rule-based model accuracy

Algorithm	Overall accuracy	Known words	Unknown words
Baseline	57.5%	57.5%	57.5%
Rule-based	65.7%	71.6%	45.2%

The system outperforms the baseline algorithm only by 8.2%, and these results are undoubtedly poor and not promising enough to continue the development of a rule-based system.

5   Neural Network Approach

The main problem Early Irish poses to machine learning methods is that its morphological complexity implies too many possible lemma classes, which, in addition to that, cannot always be reduced to a combination of a stem type and a suffix. Therefore some statistical models popular in sequence tagging that involves multi-class classification, such as HMM, MaxEnt, MEMM, SVM or CRF, are quite useless for this task. The best solution here seems to be turning from statistical machine learning to deep learning and using a sequence-to-sequence model, which allows going down to the character level. Basically, a sequence-to-sequence model is an ensemble of recurrent neural networks, or RNNs, that takes a sequence of a dynamic length as input and produces another sequence of a dynamic length. Sequence-to-sequence networks are used for a wide variety of tasks, such as grapheme-to-phoneme encoding [49], OCR post-processing, spelling correction, lemmatisation [41], machine translation [2,7] and even dialogue systems development [48]. Thus, if we reformulate the lemmatisation task as taking a sequence of characters (form) as input and generating

Table 5. Rule-based lemmatiser performance: details

Tokens	840
Known words	654
Unknown words	186
Lemmatised correctly	552
Predicted lemmas	157
Failed to predict	29
Predicted correctly	84
Predicted incorrectly	68
Disambiguation mistakes	73

another sequence of characters (lemma), we can forget about tens of verbal and nominal inflection classes, let alone spelling variation.

The data used in this experiment consists of 83,155 unique form-lemma pairs from the electronic edition of the DIL [45], shuffled and split into training, validation and test sets, the former two being 5,000 samples each. All experiments were run on a personal laptop with Intel Core i7 2,5 GHz processor and 12 Gb RAM, which took about 36 h each.

A character-to-character model was trained during 34,000 iterations, but reached minimum loss and maximum accuracy of 69.8% on a validation set after 10,000 iterations. When the training set accuracy reached its maximum, the validation set accuracy dropped to 64.9%; on the test set the model achieved 63.9%, as shown in Fig. 1.

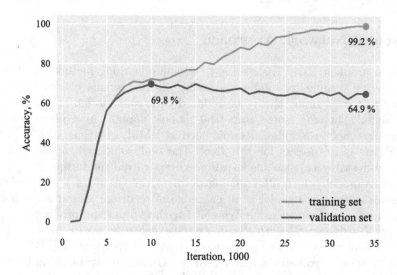

Fig. 1. Character-to-character model accuracy

These results are a serious improvement over the rule-based model, which showed only 45.2% on unknown words. Dots on accuracy graphs represent maximums on known (training set) and unknown (validation set) forms.

Having a closer look some mistakes in Table 6, made by the character-to-character model in its best configuration (further referred as *char2char*), we can clearly see that it learned to demutate forms (cf. the last two examples), but some inflection models are still unknown to it, which can be explained by the lack of training data. The model experiences most difficulties with compound verbs, which is not surprising.

Table 6. Character-to-character model mistakes

Form	Real lemma	Predicted lemma
ar-com-icc	ar-cóemsat	ar-coimcin
dáirfiniu	dáirine	dáirfinu
folortadh	folortad	folortaid
fris-tasgat	fris-tasgat	fris-taig
ithear	ithir	íthra
n-etarcnaigedar	etargnaigidir	etarncaigedar
t-iarrath	íarrath	dírarth

As poor as the results may seem, they are not very different from those achieved by sequence-to-sequence models on analogous tasks. For example, the best results for the OCR post-correction and spelling correction tasks according to [41] fall between 62.75% and 74.67% on different datasets. The score is even lower for grapheme-to-phoneme task, 44.74%–72.23% [41]. Lemmatisation scores described in the article are much higher, 94.22% for German verbs and 94.08% for Finnish verbs [41], but taking the inflectional diversity and abundant orthographic variation of Early Irish into account, this task is closer to spelling correction and grapheme-to-phoneme translation rather than to lemmatisation of any modern language. In any case, a character-level sequence-to-sequence model reached the accuracy score of 99.2% for known words and 64.9% for unknown words on a rather small corpus of 83,155 samples, which is a serious improvement over the rule-based model described above. Table 7 shows the performance of different models on Early Irish data.

The model also meets the results of other systems working with historical data. Table 8 provides a summary of best accuracy scores achieved by Early Irish, Middle Dutch [25], Latin [31] and Old French [44] lemmatisers having different architectures. Unfortunately, it is not possible to cite more results as there are no clear figures in other works concerning lemmatisation for ancient languages.

Table 7. Performance of different models on Early Irish data

Model	Accuracy (unknown)	Accuracy (known)
Baseline	57.5%	57.5%
Rule-based	45.2%	71.6%
Char2char	64.9%	99.2%

Table 8. Best accuracy scores on historical language data

Language	Model	Unknown	Known
Early Irish	Character-level seq2seq	64.9%	99.2%
Middle Dutch	CNN + word embeddings	59.48%	97.89%
Latin	CRF	81.84%	95.58%
Old French	Rule-based	?	60%

6 Conclusion

Although the task of lemmatisation for Early Irish data is quite challenging, there is a number of promising solutions. A character-level sequence-to-sequence model appears to be the best one for the moment, reaching the accuracy score of 99.2% for known words and 64.9% for unknown words on a rather small corpus of 83,155 samples. It outperforms both the baseline and the rule-based model and meets the results of other systems working with historical data.

Nevertheless, there is still much space for improvement and further research, and the first priority task that could help to ameliorate the performance is creating an open-source searchable corpus of Early Irish. It is also important to develop a detailed sensible grammatical notation to avoid such things as dropping out infixed pronouns when lemmatising verbal forms that persist in the DIL.

References

1. Attia, M., Samih, Y., Shaalan, K.F., Van Genabith, J.: The floating Arabic dictionary: an automatic method for updating a lexical database through the detection and lemmatization of unknown words. In: COLING, pp. 83–96 (2012)
2. Bahdanau, D., Cho, K., Bengio, Y.: Neural machine translation by jointly learning to align and translate. arXiv preprint (2014). arXiv:1409.0473
3. Baron, A., Rayson, P.: VARD2: a tool for dealing with spelling variation in historical corpora. In: Postgraduate Conference in Corpus Linguistics (2008)
4. Baron, A., Rayson, P.: Automatic standartisation of texts containing spelling variation. How much training data do you need (2009)
5. Bollmann, M., Dipper, S., Krasselt, J., Petran, F.: Manual and semi-automatic normalization of historical spelling-case studies from Early New High German. In: KONVENS, pp. 342–350 (2012)

6. Borin, L., Forsberg, M.: Something old, something new: a computational morphological description of Old Swedish. In: LREC 2008 Workshop on Language Technology for Cultural Heritage Data (LaTeCH 2008), pp. 9–16 (2008)

7. Cho, K., et al.: Learning phrase representations using RNN encoder-decoder for statistical machine translation. arXiv preprint (2014). arXiv:1406.1078

8. Chrupała, G.: Simple data-driven context sensitive lemmatization. Procesamiento del Leng. Nat. **37**, 121–127 (2006)

9. Chrupała, G.: Normalizing tweets with edit scripts and recurrent neural embeddings. In: Proceedings of the 52nd Annual Meeting of the Association for Computational Linguistics. vol. 2, pp. 680–686. Citeseer (2014)

10. Chrupała, G., Dinu, G., Van Genabith, J.: Learning morphology with Morfette (2008)

11. Cinková, S., Pomikálek, J.: LEMPAS: a make-do lemmatizer for the Swedish PAROLE-corpus. Prague Bull. Math. Linguist. **86**, 47–54 (2006)

12. Daelemans, W., Groenewald, H.J., Van Huyssteen, G.B.: Prototype-based active learning for lemmatization (2009)

13. De Pauw, G., De Schryver, G.M.: Improving the computational morphological analysis of a Swahili corpus for lexicographic purposes. Lexikos **18**(1), 303–318 (2008)

14. Dhonnchadha, E.U.: A two-level morphological analyser and generator for Irish using finite-state transducers. In: LREC (2002)

15. El-Shishtawy, T., El-Ghannam, F.: An accurate Arabic root-based lemmatizer for information retrieval purposes. arXiv preprint (2012). arXiv:1203.3584

16. Ernst-Gerlach, A., Fuhr, N.: Retrieval in text collections with historic spelling using linguistic and spelling variants. In: Proceedings of the 7th ACM/IEEE-CS Joint Conference on Digital Libraries, pp. 333–341. ACM (2007)

17. Giusti, R., Candido, A., Muniz, M., Cucatto, L., Aluísio, S.: Automatic detection of spelling variation in historical corpus. In: Proceedings of the Corpus Linguistics Conference (CL) (2007)

18. Halácsy, P., Trón, V.: Benefits of deep NLP-based lemmatization for information retrieval. CLEF (Working Notes) (2006)

19. Hendrickx, I., Marquilhas, R.: From old texts to modern spellings: an experiment in automatic normalisation. JLCL **26**(2), 65–76 (2011)

20. Huet, G.: Towards computational processing of Sanskrit. In: International Conference on Natural Language Processing (ICON). Citeseer (2003)

21. Ingason, A.K., Helgadóttir, S., Loftsson, H., Rögnvaldsson, E.: A mixed method lemmatization algorithm using a hierarchy of linguistic identities (HOLI). In: Nordström, B., Ranta, A. (eds.) GoTAL 2008. LNCS (LNAI), vol. 5221, pp. 205–216. Springer, Heidelberg (2008). https://doi.org/10.1007/978-3-540-85287-2_20

22. Johnson, K.P., et al.: CLTK: the classical language toolkit. https://github.com/cltk/cltk (2014–2017)

23. Kanis, J., Müller, L.: Automatic lemmatizer construction with focus on OOV words lemmatization. In: Matoušek, V., Mautner, P., Pavelka, T. (eds.) TSD 2005. LNCS (LNAI), vol. 3658, pp. 132–139. Springer, Heidelberg (2005). https://doi.org/10.1007/11551874_17

24. Kestemont, M., Daelemans, W., De Pauw, G.: Weigh your words–memory-based lemmatization for Middle Dutch. Lit. Linguist. Comput. **25**(3), 287–301 (2010)

25. Kestemont, M., de Pauw, G., van Nie, R., Daelemans, W.: Lemmatization for variation-rich languages using deep learning. Dig. Scholarsh. Humanit. **32**, 1–19 (2016)

26. Kilgarriff, A., Rundell, M., Dhonnchadha, E.U.: Efficient corpus development for lexicography: building the New Corpus for Ireland. Lang. Resour. Eval. **40**(2), 127–152 (2006)
27. Lash, E.: The parsed Old and Middle Irish corpus (POMIC). version 0.1 (2014)
28. Lyras, D.P., Sgarbas, K.N., Fakotakis, N.D.: Applying similarity measures for automatic lemmatization: a case study for Modern Greek and English. Int. J. Artif. Intell. Tools **17**(05), 1043–1064 (2008)
29. Marinone, N.: A project for Latin lexicography: 1. Automatic lemmatization and word-list. Comput. Humanit. **24**(5), 417–420 (1990)
30. Meelen, M., Beekhuizen, B.: PoS-tagging and chunking historical Welsh. In: Proceedings of the Scottish Celtic Colloquium 2012 (2013)
31. Müller, T., Cotterell, R., Fraser, A.M., Schütze, H.: Joint lemmatization and morphological tagging with Lemming. In: EMNLP, pp. 2268–2274 (2015)
32. Packard, D.: Computer-assisted morphological analysis of ancient Greek (1973)
33. Passarotti, M.C.: Development and perspectives of the Latin morphological analyser LEMLAT. Linguist. Comput. **20**(A), 397–414 (2004)
34. Paul, S., Joshi, N., Mathur, I.: Development of a Hindi lemmatizer. arXiv preprint (2013). arXiv:1305.6211
35. Perera, P., Witte, R.: A self-learning context-aware lemmatizer for German. In: Proceedings of the conference on Human Language Technology and Empirical Methods in Natural Language Processing, pp. 636–643. Association for Computational Linguistics (2005)
36. Pilz, T., Ernst-Gerlach, A., Kempken, S., Rayson, P., Archer, D.: The identification of spelling variants in English and German historical texts: manual or automatic? Lit. Linguist. Comput. **23**(1), 65–72 (2008)
37. Piotrowski, M.: Natural language processing for historical texts. Synth. Lect. Hum. Lang. Technol. **5**(2), 1–157 (2012)
38. Plisson, J., Lavrac, N., Mladenic, D., et al.: A rule based approach to word lemmatization. In: Proceedings C of the 7th International Multi-Conference Information Society IS 2004, vol. 1, pp. 83–86. Citeseer (2004)
39. Reynaert, M., Hendrickx, I., Marquilhas, R.: Historical spelling normalization. A comparison of two statistical methods: TICCL and VARD2. In: Proceedings of Annotation of Corpora for Research in the Humanities (ACRH-2), p. 87 (2012)
40. Scherrer, Y., Erjavec, T.: Modernizing historical Slovene words with character-based SMT. In: BSNLP 2013–4th Biennial Workshop on Balto-Slavic Natural Language Processing (2013)
41. Schnober, C., Eger, S., Dinh, E.L.D., Gurevych, I.: Still not there? Comparing traditional sequence-to-sequence models to encoder-decoder neural networks on monotone string translation tasks. In: Proceedings of the 26th International Conference on Computational Linguistics (COLING), December 2016, to appear
42. Segalovich, I.: A fast morphological algorithm with unknown word guessing induced by a dictionary for a web search engine. In: MLMTA, pp. 273–280. Citeseer (2003)
43. Shavrina, T., Sorokin, A.: Modeling advanced lemmatization for Russian language using TnT-Russian morphological parser. In: Computational Linguistics and Intellectual Technologies: Proceedings of the International Conference "Dialog" (2015)
44. Souvay, G., Pierrel, J.M.: Lemmatisation des mots en Moyen Français. Traitement Autom. Lang. **50**(2), 21 (2009)
45. Toner, G., Bondarenko, G., Fomin, M., Torma, T.: An electronic dictionary of the Irish language (2007)

46. Toutanova, K., Cherry, C.: A global model for joint lemmatization and part-of-speech prediction. In: Proceedings of the Joint Conference of the 47th Annual Meeting of the ACL and the 4th International Joint Conference on Natural Language Processing of the AFNLP, vol. 1, pp. 486–494. Association for Computational Linguistics (2009)
47. Verboom, A.: Towards a Sanskrit wordparser. Lit. Linguist. Comput. **3**(1), 40–44 (1988)
48. Vinyals, O., Le, Q.: A neural conversational model. arXiv preprint (2015). arXiv:1506.05869
49. Yao, K., Zweig, G.: Sequence-to-sequence neural net models for grapheme-to-phoneme conversion. arXiv preprint (2015). arXiv:1506.00196
50. Zaliznyak, A.A.: Grammatichesky slovar russkogo yazyka. Slovoizmenenie. Russian grammatical dictionary. Inflection (1980)

Named Entity Recognition in Russian with Word Representation Learned by a Bidirectional Language Model

Georgy Konoplich[1], Evgeniy Putin[1], Andrey Filchenkov[1(✉)], and Roman Rybka[2]

[1] ITMO University, Saint Petersburg, Russia
konoplich@rain.ifmo.ru, putin.evgeny@gmail.com,
afilchenkov@corp.ifmo.ru
[2] Kurchatov Institute, Moscow, Russia
RybkaRB@gmail.com

Abstract. Named Entity Recognition is one of the most popular tasks of the natural language processing. Pre-trained word embeddings learned from unlabeled text have become a standard component of neural network architectures for natural language processing tasks. However, in most cases, a recurrent network that operates on word-level representations to produce context sensitive representations is trained on relatively few labeled data. Also, there are many difficulties in processing Russian language. In this paper, we present a semi-supervised approach for adding deep contextualized word representation that models both complex characteristics of word usage (e.g., syntax and semantics), and how these usages vary across linguistic contexts (i.e., to model polysemy). Here word vectors are learned functions of the internal states of a deep bidirectional language model, which is pretrained on a large text corpus. We show that these representations can be easily added to existing models and be combined with other word representation features. We evaluate our model on FactRuEval-2016 dataset for named entity recognition in Russian and achieve state of the art results.

Keywords: NER · Word representation · Semi-supervised learning
Language modeling · Bi-LSTM

1 Introduction

Due to their simplicity and efficiency, pre-trained word embedding have become widespread in natural language processing (NLP) systems. Many prior studies have shown that such embedding capture useful semantic and syntactic information [1, 2] and including them in NLP systems has been shown to be highly helpful for a variety of domain tasks [3]. However, these approaches for learning word vectors only allow a single context independent representation for each word. Learning high quality representations can be challenging.

Previously proposed methods overcome some of the shortcomings of traditional word vectors by either enriching them with subword information [4, 5] or learning separate vectors for each word sense [6]. Other recent work has also focused on learning

© Springer Nature Switzerland AG 2018
D. Ustalov et al. (Eds.): AINL 2018, CCIS 930, pp. 48–58, 2018.
https://doi.org/10.1007/978-3-030-01204-5_5

context-dependent representations. Context2vec [7] uses a bidirectional Long Short Term Memory (LSTM) to encode the context around a pivot word. Other approaches for learning contextual embeddings include the pivot word itself in the representation and are computed with the encoder of either a supervised neural machine translation (MT) system [8] or an unsupervised language model [9]. Both approaches benefit from large datasets, although the MT approach is limited by the size of parallel corpora.

Previous work has also shown that different layers of deep bidirectional recurrent neural networks (biRNNs) encode different types of information. For example, introducing multi-task syntactic supervision (e.g., part-of-speech tags) at lower levels of a deep LSTM can improve overall performance of higher level tasks such as dependency parsing [10]. Authors of [11] showed that in an RNN-based encoder-decoder machine translation system, the representations learned at the first layer in a 2-layer LSTM encoder are better at predicting POS tags than the representations learned at the second layer. Finally, the top layer of an LSTM for encoding word has been shown to learn representations of word sense [7].

State of the art sequence tagging models typically include a biRNN that encodes word sequences into a context sensitive representation before making word specific predictions [10, 12, 13]. The problem with these models is that they are trained on a small amount of labeled data and do not fully take into account the context of each word. Authors of [12] have presented methods for jointly learning the biRNN with supplemental labeled data from other tasks.

There are many difficulties in processing the Russian language: free word order, morphological richness, polysemy, neologisms. The approach we suggest aims to handle with all these difficulties.

In this paper, we explore an alternate semi-supervised approach, which does not require additional labeled data. We use Embeddings from Language Models (ELMo) representations [14] (Sect. 2.3). Unlike previous approaches for learning contextualized word vectors [8, 9], ELMo representations are deep, in the sense that they are a function of all internal layers of the deep bidirectional language model (biLM, Sect. 2.1). The biLM architecture is described in Sect. 2.4. Our biLM train on large corpus of unlabeled Russian data, then we use fine-tuning of biLM model on task specific data, supervised labels are temporarily ignored. For fine-tuning we use approach similar to the one presented in paper [15]. We use discriminative fine-tuning and gradual unfreezing, techniques to retain previous knowledge and avoid catastrophic forgetting during fine-tuning (Sect. 2.2). Then combined fastText word representation and ELMo embeddings are given to a bidirectional LSTM (Sect. 3).

2 Word Representation

Unlike the most widely used word embeddings [1, 2], ELMo word representations are functions of the entire input sentence. They are computed on the top of two-layer biLMs with character convolutions, as a linear function of the internal network states. This setup allows to perform semi-supervised learning, where the biLM is pretrained at a large scale and incorporated into a wide range of existing neural NLP architectures. We also fine-tune the biLM on domain specific data to increase performance of the model for NER.

2.1 Bidirectional Language Models

Given a sequence of N tokens, $(t_1, t_2, \ldots, t_N)$, a forward language model computes the probability of the sequence by modeling the probability of token t_k given the history $(t_1, \ldots, t_{k-1})$:

$$p(t_1, t_2, \ldots, t_N) = \prod_{k=1}^{N} p(t_k | t_1, t_2, \ldots, t_{k-1}).$$

Recent state-of-the-art neural language models [16] compute a context-independent token representation x_k^{LM} (via token embeddings or a CNN over characters) then pass it through L layers of forward LSTMs. At each position k, each LSTM layer outputs a context-dependent representation $\overrightarrow{h_{k,j}^{LM}}$, where $j = 1, \ldots, L$. The top layer LSTM output $\overrightarrow{h_{k,L}^{LM}}$ is used to predict the next token t_{k+1} with a Softmax layer. A backward LM is similar to a forward LM, except it runs over the sequence in reverse, predicting the previous token given the future context. It can be implemented in an analogous way to a forward LM, with each backward LSTM layer j in a L layer deep model producing representations $\overleftarrow{h_{k,j}^{LM}}$ of t_k given $(t_{k+1}, \ldots, t_N)$.

A biLM combines both the forward and backward LM. This formulation jointly maximizes the log likelihood of the forward and backward directions [14]:

$$\sum_{k=1}^{N} \left(\log p(t_k | t_1, \ldots, t_{k-1}; \Theta_x, \overrightarrow{\Theta_{LSTM}}, \Theta_S) + \log p(t_k | t_{k+1}, \ldots, t_N; \Theta_x, \overleftarrow{\Theta_{LSTM}}, \Theta_s) \right).$$

The parameters tied for both the token representation (Θ_x) and Softmax layer (Θ_s) in the forward and backward direction while maintaining separate parameters for the LSTMs in each direction. Overall, this formulation is similar to the approach presented in [9], with the exception that some weights shared between directions instead of using completely independent parameters.

2.2 Fine-Tuning BiLM

For fine-tuning, we use approach similar to the one presented in [15]. We use discriminative fine-tuning and gradual unfreezing techniques to retain previous knowledge and avoid catastrophic forgetting during fine-tuning.

Discriminative Fine-Tuning. As different layers capture different types of information [17], they should be fine-tuned to different extents. To this end, we use discriminative fine-tuning. Instead of using the same learning rate for all layers of the model, discriminative fine-tuning allows us to tune each layer with different learning rates. For context, the regular stochastic gradient descent (SGD) update of a model parameters θ at time step t looks like the following [18]:

$$\theta_t = \theta_{t-1} - \eta \cdot \nabla_\theta J(\theta),$$

where η is the learning rate and $\Delta_\theta J(\theta)$ is the gradient with regard to the model objective function. For discriminative fine-tuning, we split the parameters θ into $\{\theta^1, ..., \theta^L\}$, where θ^l contains the parameters of the model at the l-th layer and L is the number of layers of the model. Similarly, we obtain $\{\eta_1, ..., \eta_L\}$ where η_l is the learning rate of the l-th layer.

The SGD update with discriminative finetuning is then the following:

$$\theta_t^l = \theta_{t-1}^l - \eta^l \cdot \nabla_{\theta^l} J(\theta).$$

We choose learning rate η_L of the last layer. The learning rate for lower layers is chosen as $\eta^{l-1} = \frac{\eta^l}{2.6}$ guided by an empirical research.

Gradual Unfreezing. Rather than fine-tuning all layers at once, which risks catastrophic forgetting, we use to gradually unfreeze the model starting from the last layer as this contains the least general knowledge [17]. We first unfreeze the last layer and finetune all unfrozen layers for one epoch. We then unfreeze the next lower frozen layer and repeat, until we finetune all layers until convergence at the last iteration. This is similar to "chain-thaw" [19], except that we add a layer at a time to the set of "thawed" layers, rather than only training a single layer at a time.

2.3   ELMo Embeddings

ELMo is a task specific combination of the intermediate layer representations in the biLM. For each token t_k, a L-layer biLM computes a set of $2L + 1$ representations

$$R_k = \{x_k^{LM}, \overrightarrow{h_{k,j}^{LM}}, \overleftarrow{h_{k,j}^{LM}} | j = 1, ..., L\} = \{h_{k,j}^{LM} | j = 0, ..., L\},$$

where $h_{k,0}^{LM}$ is the token layer and $h_{k,j}^{LM} = \left[\overrightarrow{h_{k,j}^{LM}}; \overleftarrow{h_{k,j}^{LM}} \right]$ for each biLSTM layer. For inclusion in a downstream model, ELMo collapses all layers in R into a single vector, $ELMo_k = E(R_k; \Theta_e)$. In the simplest case, ELMo just selects the top layer, $E(R_k) = h_{k,L}^{LM}$, as in TagLM [9] and CoVe [8]. More generally, word representation using all biLM layers computed as:

$$ELMo_k = E(R_k; \Theta) = \gamma \cdot \sum_{j=0}^{L} s_j h_{k,j}^{LM},$$

where, s are softmax-normalized weights and the scalar parameter γ allows the task model to scale the entire ELMo vector. Parameter γ is of practical importance to aid the optimization process. Considering that the activations of each biLM layer have a different distribution, in some cases it also helped to apply layer normalization [20] to each biLM layer before weighting.

2.4   Pre-trained Bidirectional Language Model Architecture

In this paper, we use following BiLM AllenNLP TensorFlow implementation[1]. This BiLM architecture is similar to the architectures described in [16], but modified to support joint training of both directions and add a residual connection between LSTM layers (Fig. 1).

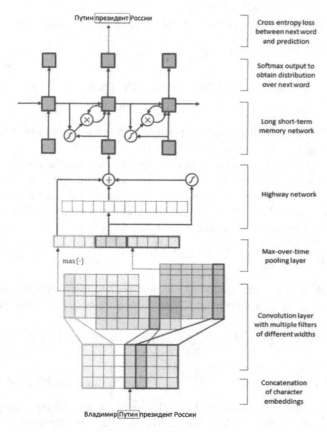

Fig. 1. BiLM architecture

To balance overall language model perplexity with the model size and computational requirements for downstream tasks while maintaining a purely character-based input representation, we change all embedding and hidden dimensions from the single best model CNN-BIG-LSTM in [16]. The final model uses $L = 2$ biLSTM layers with 2048 units and 512 dimension projections and a residual connection from the first to the second layer. The context insensitive type representation uses 2048 character n-gram convolutional filters followed by two highway layers [21] and a linear projection down to a 512 representation.

---

[1] https://github.com/allenai/bilm-tf.

3   Bidirectional LSTM + CRF Model for NER

Following recent state-of-the-art systems [9, 12], the baseline model uses pre-trained word embeddings, two biLSTM layers and a conditional random field (CRF) loss [22], similar to [3]. The different word representation is passed through two biLSTM layers, the first with 40 hidden units and the second with 20 hidden units before a final dense layer. During training, we use a CRF loss and at test time perform decoding using the Viterbi algorithm. The architecture of the model is presented on Fig. 2.

The implementation of the model can be found in online repository[2].

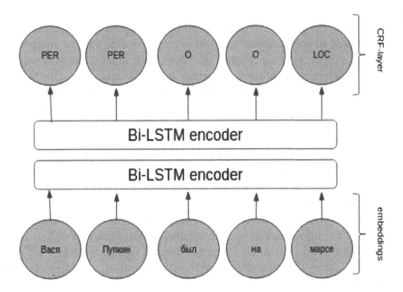

Fig. 2. The main network architecture for NER. Word embeddings are given to a bidirectional LSTM. Example of input is in Russian: "Vasya Pupkin was on Mars"

4   Experiments and Results

4.1   Dataset and Evaluation

We use the FactRuEval corpus of data[3]. The corpus consists of newswire and analytical texts in Russian dealing with social and political issues. The texts were gathered from Private Correspondent[4] and Wikinews[5].

---

[2] https://github.com/ctlab/ML/BiLSTM_for_NER.

[3] https://github.com/dialogue-evaluation/factRuEval-2016.

[4] http://www.chaskor.ru/.

[5] https://ru.wikinews.org.

The corpus is split into two parts—a demo corpus of 122 texts and a test corpus of 133 texts. The demo corpus contains about 30 thousand tokens and 1700 sentences. The test corpus contains about 60 thousand tokens and 3100 sentences. We train our models on the demo corpus and test the on the test corpus. Entities of the following three types have to be recognized: person, location, organization. We use official FactRuEval-2016 evaluation scripts to calculate metrics of performance (micro-averaged F_1).

4.2 Baseline

As a baseline solution, we tried to train a neural network only on morphological and syntactic features.

For word representation, we use the information contained in the syntax tree, the surrounding words in the sentence with the window size 5, various morphological features: prefix, postfix, post-tag, lemma and others. The input dimension was 825.

We tried different architectures of neural networks and used GridSearch for hyperparameter optimization. The best architecture of neural network consists of three recurrent layers (LSTM), two dense layers and multi-layer perceptron at the end. The architecture of the model is presented on the Fig. 3.

Full set of parameters for this model consists of parameters of neural network layers (weight matrices, biases, word embedding matrix). All these parameters are tuned during training stage by back propagation algorithm with stochastic gradient descent ($lr = 0.1, decay = 10^{-7}, momentum = 0, clipvalue = 3$). Variational Dropout is applied to avoid overfitting and to improve the system performance.

After training 30 epochs, RNN + MLP model achieves 72.90 *precision*, 76.60 *recall* and 74.71 F_1. Input features encode insufficient knowledge of natural language and the model is trained on a small amount of labeled data. It requires a lot of word knowledge, and a deep understanding of grammar, semantics, and other elements of natural language.

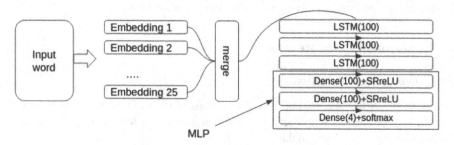

Fig. 3. Baseline solution. Input word representation combine syntactic and morphological features and feed it to 3-layers LSTM model with MLP at the end.

Then for improving results, we add distributive word representation for each word. However, word representations only give a single context independent representation for each word. And at the end we will show how Embeddings from Language Models

(ELMo) improve results. ELMo give a context dependent representation for each word and solve the problem of polysemy.

4.3   Training BiLM Model and Fine-Tuning

We train biLM model on large corpus of data. Data are taken from conll2017[6]. After preprocessing, we have about 200 million tokens and 15 million sentences, the vocabulary is 60 thousand most frequent words. After training for 5 epochs, the average forward and backward perplexities is about 47. Once pretrained, the biLM can compute representations for any task.

In some cases, fine-tuning the biLM on domain specific data leads to significant drops in perplexity and an increase in downstream task performance. This can be seen as a type of domain transfer for the biLM. To fine-tune the model, we take raw sentences of demo corpus. The biLM is fine-tuned as described in Subsect. 2.2 for 3 epochs and evaluated on the test corpus sentences. Test corpus perplexity after fine-tuning improved from 90 to 40 (lower is better). Once fine-tuned, the biLM weights were fixed during task training.

4.4   Training Bi-LSTM + CRF

FastText word representations and some morphological features are given to model. Full set of parameters for this model consists of parameters of Bi-LSTM layers (weight matrices, biases, word embedding matrix) and transition matrix of CRF layer. All these parameters are tuned during training stage by back propagation algorithm with stochastic gradient descent ($lr = 0.1, decay = 10^{-7}, momentum = 0, clipvalue = 3$). Variational Dropout is applied to avoid over-fitting and improve the system performance. After training for 30 epochs, Bi-LSTM + CRF model achieves 77.23 *precision*, 85.19 *recall* and 81.02 F_1.

Then we combine FastText word representations and ELMo embeddings before fine tuning and repeated training model. After training for 30 epochs, Bi-LSTM + CRF model achieves 82.32 *precision*, 84.04 *recall* and 83.17 F_1.

Finally, we fine-tune biLM model and take ELMo embeddings. After training for 30 epochs, Bi-LSTM + CRF model achieves 83.19 *precision*, 85.41 *recall* and 84.29 F_1.

4.5   Results and Other Works

Traditional approaches to named entity recognition in Russian are based on hand-crafted rules and external resources. So, in work [23], regular expressions and dictionaries were used to solve the problem. The next step was the application of statistical training methods, such as conditional random fields (CRF) and supporting vector machines (SVM) for classifying entities. CRF over linguistic features, considered as the baseline in the studies [24]. In [25], a two-stage algorithm of conditional random fields was proposed: at the first stage, the input for the CRF was on hand-crafted linguistic

---

[6] http://universaldependencies.org/conll17/.

features. Then on the second, the same linguistic features were combined with global statistics, calculated at the first stage and submitted to the CRF. In work [26], SVM was applied to the distributive vector representations of words and phrases. These representations were obtained by extensive unsupervised pre-training on different news corpora. Simultaneous use of dictionary-based features and distributed word representations was presented in [27]. Dictionary features were retrieved from Wikidata and word representations were pre-trained on Wikipedia. Then these features were used for classification with SVM.

In [28], LSTM networks were applied to NER task in Russian. In the study [29], a modern model of the neural network for the English NER is applied to open data mappings for NER in Russian. The model consists of three main components: bidirectional recurrent networks (bi-LSTM), CRF and distributed vector representation of words. Such topology is widely used in for sequential data processing in different domains [30, 31]. A deeper study on how these models can be combined was presented in [32].

Table 1 presents the results of studies in which the evaluating phase corresponds to the requirements of the competition factRuEval-2016. We compare our results with methods, proposed in [27, 29, 32] as showing state-of-the-art results. We did not include [26], because the authors achieved the results on strings that were known to contain NER term, while we solve the problem without such knowledge, for entire text, some strings of which may not contain any NER term.

Table 1. Comparison of different models.

Model	Precision	Recall	F1-score
Rubaylo and Kosenko [29]	77.70	78.50	78.13
Basic + dictionary + w2v features + SVM [27]	82.57	74.08	78.10
Bi-LSTM + CRF + Lenta [32]	**83.8**	80.84	82.10
NeuroNER + Highway char [32]	80.59	80.72	80.66
RNN + MLP	72.90	76.60	74.71
Bi-LSTM-CRF	77.23	85.19	81.02
Bi-LSTM-CRF + ELMo	82.32	84.04	83.17
Bi-LSTM-CRF + ELMo + fine-tuning	83.19	**85.41**	**84.29**

As it can be seen, usage of ELMo helped to increase model performance dramatically in terms of precision at the cost of decreasing its recall. Fine-tuning increased both precision and recall even greater, resulting in the best model we used in comparison.

5 Conclusions

In this work, we showed that the context sensitive representation captured in the ELMo embeddings is useful in named entity recognition in Russian. When we fine-tuned our bidirectional language model and included ELMo embeddings in our Bi-LSTM, we achieve state-of-the-art results.

There is no clear understanding between the performance of the language model and the accuracy in solving specific domain tasks. It is not clear when the language model has learned better and its application will yield better results on the inherited model. Further it is supposed to continue studying language models. Training them in the Russian language, experiments with different architectures, obtaining deep context-dependent representations. After all, such representations are useful in many tasks of natural language processing.

Acknowledgements. The authors would like to thank Ivan Smetannikov for helpful conversation and unknown AINL reviewers for useful comments.

The research was supported by the Government of the Russian Federation (Grant 08-08).

References

1. Mikolov, T., Sutskever, I., Chen, K., Corrado, G.S., Dean, J.: Distributed representations of words and phrases and their compositionality. In: Advances in Neural Information Processing Systems, pp. 3111–3119 (2013)
2. Pennington, J., Socher, R., Manning, C.: GloVe: global vectors for word representation. In: Proceedings of the 2014 Conference on Empirical Methods in Natural Language Processing (EMNLP), pp. 1532–1543 (2014)
3. Collobert, R., Weston, J., Bottou, L., Karlen, M., Kavukcuoglu, K., Kuksa, P.: Natural language processing (almost) from scratch. J. Mach. Learn. Res. **12**, 2493–2537 (2011)
4. Wieting, J., Bansal, M., Gimpel, K., Livescu, K.: Charagram: embedding words and sentences via character n-Grams. arXiv:1607.02789 (2016)
5. Bojanowski, P., Grave, E., Joulin, A., Mikolov, T.: Enriching word vectors with subword information. arXiv:1607.04606 (2016)
6. Neelakantan, A., Shankar, J., Passos, A., McCallum, A.: Efficient non-parametric estimation of multiple embeddings per word in vector space. arXiv:1504.06654 (2015)
7. Melamud, O., Goldberger, J., Dagan, I.: context2vec: learning generic context embedding with bidirectional LSTM. In: Proceedings of the 20th SIGNLL Conference on Computational Natural Language Learning, pp. 51–61 (2016)
8. McCann, B., Bradbury, J., Xiong, C., Socher, R.: Learned in translation: contextualized word vectors. In: Advances in Neural Information Processing Systems, pp. 6297–6308 (2017)
9. Peters, M.E., Ammar, W., Bhagavatula, C., Power, R.: Semi-supervised sequence tagging with bidirectional language models. arXiv:1705.00108 (2017)
10. Hashimoto, K., Xiong, C., Tsuruoka, Y., Socher, R.: A joint many-task model: growing a neural network for multiple NLP tasks. arXiv:1611.01587 (2016)
11. Belinkov, Y., Durrani, N., Dalvi, F., Sajjad, H., Glass, J.: What do neural machine translation models learn about morphology? arXiv:1704.03471 (2017)
12. Yang, Z., Salakhutdinov, R., Cohen, W.W.: Transfer learning for sequence tagging with hierarchical recurrent networks. arXiv:1703.06345 (2017)
13. Lample, G., Ballesteros, M., Subramanian, S., Kawakami, K., Dyer, C.: Neural architectures for named entity recognition. arXiv:1603.01360 (2016)
14. Peters, M.E., et al.: Deep contextualized word representations. arXiv:1802.05365 (2018)
15. Howard, J., Sebastian, R.: Fine-tuned language models for text classification. arXiv:1801.06146 (2018)

16. Jozefowicz, R., Vinyals, O., Schuster, M., Shazeer, N., Wu, Y.: Exploring the limits of language modeling. arXiv:1602.02410 (2016)
17. Yosinski, J., Clune, J., Bengio, Y., Lipson, H.: How transferable are features in deep neural networks? In: Advances in Neural Information Processing Systems, pp. 3320–3328 (2014)
18. Ruder, S.: An Overview of gradient descent optimization algorithms. arXiv:1609.04747 (2016)
19. Felbo, B., Mislove, A., Søgaard, A., Rahwan, I., Lehmann, S.: Using millions of Emoji occurrences to learn any-domain representations for detecting sentiment, emotion and sarcasm. arXiv:1708.00524 (2017)
20. Ba, J.L., Kiros, J.R., Hinton, G.E.: Layer normalization. arXiv:1607.06450 (2016)
21. Srivastava, R.K., Greff, K., Schmidhuber, J.: Training very deep networks. In: Advances in Neural Information Processing Systems, pp. 2377–2385 (2015)
22. Lafferty, J., McCallum, A., Pereira, F.C.: Conditional random fields: probabilistic models for segmenting and labeling sequence data (2001)
23. Trofimov, I.V.: Person name recognition in news articles based on the per-sons1000/1111-F collections. In: 16th All-Russian Scientific Conference Digital Libraries: Advanced Methods and Technologies, Digital Collections, RCDL 2014, pp. 217–221 (2014)
24. Gareev, R., Tkachenko, M., Solovyev, V., Simanovsky, A., Ivanov, V.: Introducing baselines for russian named entity recognition. In: Gelbukh, A. (ed.) CICLing 2013 Part I. LNCS, vol. 7816, pp. 329–342. Springer, Heidelberg (2013). https://doi.org/10.1007/978-3-642-37247-6_27
25. Mozharova, V., Loukachevitch, N.: Two-stage approach in Russian named entity recognition. In: Proceeding of IEEE International FRUCT Conference on Intelligence, Social Media and Web (ISMW FRUCT 2016), pp. 1–6 (2016)
26. Ivanitskiy, R., Shipilo, A., Kovriguina, L.: Russian named entities recognition and classification using distributed word and phrase representations. In: SIMBig, pp. 150–156 (2016)
27. Sysoev, A.A., Andrianov, I.A.: Named entity recognition in Russian: the power of wiki-based approach. In: Dialog Conference (2016, in Russian)
28. Malykh, V., Ozerin, A.: Reproducing Russian NER baseline quality without additional data. In: CDUD@ CLA, pp. 54–59 (2016)
29. Rubaylo, A.V., Kosenko, M.Y.: Software utilities for natural language information retrievial. Alm. Mod. Sci. Educ. **12**(114), 87–92 (2016)
30. Huang, Z., Xu, W., Yu, K.: Bidirectional LSTM-CRF models for sequence tagging. arXiv: 1508.01991 (2015)
31. Tutubalina, E., Nikolenko, S.: Combination of deep recurrent neural networks and conditional random fields for extracting adverse drug reactions from user reviews. J Healthc. Eng. 2017 (2017)
32. Anh, L.T., Arkhipov, M.Y., Burtsev. M.S.: Application of a hybrid Bi-LSTM-CRF model to the task of russian named entity recognition. arXiv:1709.09686 (2017)

Sentence and Discourse Representations

Supervised Mover's Distance: A Simple Model for Sentence Comparison

Muktabh Mayank Srivastava[✉]

ParallelDots, Inc., Gurugram, India
muktabh@paralleldots.com
https://paralleldots.xyz

Abstract. We propose a simple neural network model which can learn relation between sentences by passing their representations obtained from Long Short Term Memory (LSTM) through a Relation Network. The Relation Network module tries to extract similarity between multiple contextual representations obtained from LSTM. The aim is to build a model which is simple to implement, light in terms of parameters and works across multiple supervised sentence comparison tasks. We show good results for the model on two sentence comparison datasets.

Keywords: Supervised Mover's Distance · Sentence comparison
Paraphrase detection · Natural language inference

1 Introduction

Sentence Comparison is a common NLP task which comes up in multiple domains. Sentence comparison measure might be needed to check redundant data [6] or check sentences for being paraphrases [3]. We propose a new method to compare sentences for both these tasks, which uses Relation Networks (RN) module [11] in combination with a Long Short Term Memory (LSTM) [4]. To compare two sentences, all possible pairs of dense vectors, one from each sentence in a pair, are passed through a Relation Network module to decipher relationship information between sentences. To make sure the dense vectors passed to Relation Network have contextual information, sentences are individually passed through a LSTM and the hidden units obtained for each word of a sentence are used as dense vectors. The inspiration of the model comes from Earth Mover's Distance (EMD) [10] which can be used to calculate distances between two distributions of points represented by vectors by optimal weighted comparison of points pairwise. The assumption is that LSTM can generate contextual vectors which can be then fed pairwise to RN to determine similarity. This is the reason for referring the algorithm as Supervised Mover's Distance, however, the algorithm does not solve optimal transport like EMD.

© Springer Nature Switzerland AG 2018
D. Ustalov et al. (Eds.): AINL 2018, CCIS 930, pp. 61–66, 2018.
https://doi.org/10.1007/978-3-030-01204-5_6

2   Previous Work

In our experiments, we focus on two sentence comparison tasks: 1. Duplication detection between questions [6] and 2. Paraphrase detection [3]. Duplication detection task aims to check whether two questions intend to ask about the same topic. Paraphrase Detection task aims to classify sentences according to whether they have a paraphrase/semantic equivalence relationship. Deep Neural Networks have shown state of the art performance in sentence comparison tasks. Most top methods for paraphrase detection are based on Deep Neural Networks [1,2]. BiMPM model [12] combines a custom matching layer with LSTMs [4] for question duplication detection.

Relation Networks (RN) [11] was introduced as a simple module for relational reasoning. The module has been used for spatial relational reasoning in images earlier, but we try to use it for deciphering relationships in text by combining it with an LSTM. RNs operate on a set of objects without regard to the objects' order, so we use LSTMs to extract out temporal information containing word importances and use RNs on top for reasoning. RN module has a g-layer which models relation between all possible pairs of objects and a f-layer which models the final output looking at the relation between objects.

Another set of models which use pairwise relationships to model document similarity are Word Mover's Distance (WMD) [8] and its supervised variant (SWMD) [5]. They both are methods to calculate Earth Mover's Distance (EMD) [10] between documents for document calculation. Both these methods calculate flows (weightages) to be given distances between each possible pair of words to calculate document distance. WMD is an unsupervised distance measure between documents. The SWMD architecture works on longer documents (with more than 40 words) and uses a complex optimization procedure to optimize EMD. SWMD uses a cascaded loss where the inner loss optimizes word importance and outer loss optimizes EMD flow. Our method is inspired from WMD and SWMD algorithms as it takes pairwise modelling of words into account but tries to achieve it using a single RN module. However, our method is not trying to solve optimal transport problem, but is trying to use contextual vectors derived from LSTM in pairs as input to RN module to model similarity.

3   Method

As Supervised Mover's Distance, we propose a baseline that generalizes well across different tasks. Our network combines LSTM layers [4] with a RN [11] module modeling semantic relationship between the sentences. The neural network architecture we propose is trained on pair of sentences to predict one of various classes the pair might fall into. For redundancy detection and paraphrase detection the labels are positive or negative, but might be different for any other tasks. The architecture has two basic parts: 1. LSTM layers and 2. RN layer. The LSTM layers can have depth of one or higher which take both sentences as input individually and produce hidden layers as output for each of the words in the sentences. This would yield two series of output hidden states, one hidden state for each time step of each sentence. To clarify again, there is one common LSTM which runs on both sentences separately to produce respective hidden states. In the RN, all possible pairs of hidden states across both sentences are taken as concatenated vectors and passed through a fully connected (or Dense) layer. Aforementioned fully connected layer is the g-layer of the RN. This yields an embedding for each possible pair of hidden state outputs from the LSTM. These embeddings are averaged and passed through another fully connected layer to predict the output. This fully connected layer is the f-layer of the RN. By taking all pairs of hidden states and using them to model sentence comparison task, we hypothesize that the LSTM would be able to make contextual vectors and RN can model pairwise differences to understand relationships between two sentences.

We illustrate the architecture part-by-part in upcoming Figs. 1, 2 and 3. Lets say two sentences s1 and s2 are to be compared. The sentence s1 is processed by a LSTM as shown in Fig. 1. The same LSTM processes s2 to get its hidden states as in Fig. 2. Now hidden states obtained from both s1 and s2 through LSTM are grouped into pairs (each pair has one hidden state from s1 and another from s2) and classified into possible classes using RN. The RN is potrayed in Fig. 3. Please note that although LSTM and RN are depicted in different diagrams for explanation, both the modules are part of the same neural network architecture and are backpropogated together. Our model is light in terms of parameters as it has only a LSTM layer and two dense (fully connected) layers in RN. A limiting case of the architecture can be when the number of LSTM layers is zero, and word embeddings are passed as inputs directly to RN.

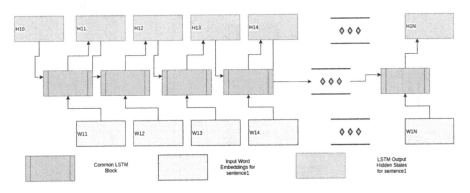

Fig. 1. Embeddings from sentence 1 are passed through an LSTM and its hidden states are taken corresponding to each input word

The network is trained with common hyperparameters for both the tasks. Pretrained word embeddings are used to initialize the word embedding layer which are finetuned by backpropogation. We use the publicly available 6 Billion token 100 dimensional version of GloVe embeddings [9]. The hidden state output from the LSTM is 100 dimensions and the size of embedding generated in the relational layers is 100 dimensions too. The network is trained with simple Stochastic Gradient Descent with momentum (common values for training across both datasets, learning rate = 0.001, momentum = 0.9).

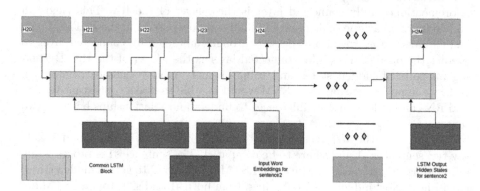

Fig. 2. The same LSTM is used to process sentence 2 and obtain each of its hidden states corresponding to its input words

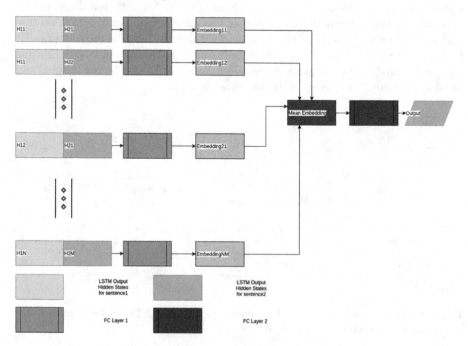

Fig. 3. A Relation Network (RN) is used to process all possible pairs of hidden states (one from both sentences which need to be compared)

4 Results

As stated we test our model on two datasets. Model is compared to state of the art methods and baselines for each dataset in this section.

Microsoft Research Paraphrase Corpus. Microsoft paraphrase corpus [3] is a corpus of sentence pairs classified as paraphrases or non-paraphrases. The dataset has 4076 sentences in training set and 1725 sentences in test set. Our model was trained on the training set with the standard set of hyper parameters mentioned above and evaluated on the test set. The accuracy numbers of different models were taken from this url[1]. Our model gets an accuracy of 80.2% on the dataset as compared to state of the art accuracy of 80.4% [7].

Quora Questions' Pair Dataset. Quora Questions' Pair Dataset contains question pairs from the Q&A website[2] tagged as similar or not. A random 90%–10% train-test split is performed as is customary for other methods and the model is trained on the train set and evaluated on the test set. As in case of other datasets, the hyperparameters are fixed as the standard values specified earlier while training. Our model gets an accuracy of 81.2% on the dataset. List of state of the art models on the dataset is available on this url[3]. The best accuracy a model gets on the dataset is 88% [12]. Although our model doesn't get results as good as the state of the art, it is competitive to baselines like siamese Convolutional Neural Networks (79.6%) and siamese LSTMs (82.58%).

It should be noted that in both models, dataset specific hyperparameter tuning was not performed.

5 Conclusion and Future Work

We propose a new method which uses a new and simple neural network model to compare sentences. Our method combines Long Short Term Memory (LSTM) and Relation Network (RN) module to model relationship between the sentences. LSTMs generate contextual hidden state vectors and RN module models sentence relationship. Models performance is calculated on two sentence comparison datasets. In future work, we will incorporate trainable components in our method to determine importance of each component as the current RN takes a mean over all vectors, treating each of the components with equal weightage.

---

[1] https://aclweb.org/aclwiki/Paraphrase_Identification_(State_of_the_art).
[2] quora.com.
[3] https://github.com/bradleypallen/keras-quora-question-pairs.

References

1. Cheng, J., Kartsaklis, D.: Syntax-aware multi-sense word embeddings for deep compositional models of meaning. arXiv preprint arXiv:1508.02354 (2015)
2. Croce, D., Moschitti, A., Basili, R.: Structured lexical similarity via convolution kernels on dependency trees. In: Proceedings of the Conference on Empirical Methods in Natural Language Processing, pp. 1034–1046. Association for Computational Linguistics (2011)
3. Dolan, B., Brockett, C.: Automatically constructing a corpus of sentential paraphrases. In: Third International Workshop on Paraphrasing (IWP2005). Asia Federation of Natural Language Processing, January 2005. https://www.microsoft.com/en-us/research/publication/automatically-constructing-a-corpus-of-sentential-paraphrases/
4. Hochreiter, S., Schmidhuber, J.: Long short-term memory. Neural Comput. **9**(8), 1735–1780 (1997)
5. Huang, G., Guo, C., Kusner, M.J., Sun, Y., Sha, F., Weinberger, K.Q.: Supervised word mover's distance. In: Advances in Neural Information Processing Systems, pp. 4862–4870 (2016)
6. Iyar, S., Dandekar, N., Csernai, K.: First quora dataset release: question pairs, January 2016. https://data.quora.com/First-Quora-Dataset-Release-Question-Pairs
7. Ji, Y., Eisenstein, J.: Discriminative improvements to distributional sentence similarity. In: Proceedings of the 2013 Conference on Empirical Methods in Natural Language Processing, pp. 891–896 (2013)
8. Kusner, M.J., Sun, Y., Kolkin, N.I., Weinberger, K.Q.: From word embeddings to document distances. In: Proceedings of the 32nd International Conference on International Conference on Machine Learning, ICML 2015, vol. 37, pp. 957–966. JMLR.org (2015). http://dl.acm.org/citation.cfm?id=3045118.3045221
9. Pennington, J., Socher, R., Manning, C.D.: GloVe: global vectors for word representation. In: Empirical Methods in Natural Language Processing (EMNLP), pp. 1532–1543 (2014). http://www.aclweb.org/anthology/D14-1162
10. Rubner, Y., Tomasi, C., Guibas, L.J.: A metric for distributions with applications to image databases. In: Sixth International Conference on Computer Vision, pp. 59–66. IEEE (1998)
11. Santoro, A., et al.: A simple neural network module for relational reasoning. In: Advances in Neural Information Processing Systems, pp. 4974–4983 (2017)
12. Wang, Z., Hamza, W., Florian, R.: Bilateral multi-perspective matching for natural language sentences. arXiv preprint arXiv:1702.03814 (2017)

Direct-Bridge Combination Scenario for Persian-Spanish Low-Resource Statistical Machine Translation

Benyamin Ahmadnia[1(✉)], Javier Serrano[1], Gholamreza Haffari[2],
and Nik-Mohammad Balouchzahi[3]

[1] Autonomous University of Barcelona, Cerdanyola del Valles, Spain
benyamin.busari@gmail.com, javier.serrano@uab.cat
[2] Monash University, Clayton, VIC, Australia
gholamreza.haffari@monash.edu
[3] University of Sistan and Baluchestan, Zahedan, Iran
balouchzahi@ece.usb.ac.ir

Abstract. This paper investigates the idea of making effective use of bridge language technique to respond to minimal parallel-resource data set bottleneck reality to improve translation quality in the case of Persian-Spanish low-resource language pair using a well-resource language such as English as the bridge one. We apply the optimized direct-bridge combination scenario to enhance the translation performance. We analyze the effects of this scenario on our case study.

Keywords: Statistical Machine Translation
Low-resource languages · Bridge language technique

1 Introduction

Since state-of-the art Statistical Machine Translation (SMT) has shown that high-quality translation output is heavily dependent on the availability of massive amounts of parallel texts in the source and target languages, the biggest issue is that high-quality parallel corpus is not always available. This is one of the reasons that SMT is to introduce a third language, named bridge (pivot) for the purpose of resolving the training data scarcity. This third language will act as an intermediary language for which there exist high-quality source-bridge and bridge-target bilingual corpora. The primary goal of SMT is to conduct the translation of the source language sequences into a target language. This must be achieved after plausibility of the source has been assessed along with the target sequences. At this point, only those target sequences must be analyzed that have a specific relation to the existing bodies of translation between the two languages [1]. Special effects are incurred by the sizable bodies of aligned parallel corpora on the functions and performance of SMT systems. However, gathering parallel data becomes quite an issue if it has to be done in practice

© Springer Nature Switzerland AG 2018
D. Ustalov et al. (Eds.): AINL 2018, CCIS 930, pp. 67–78, 2018.
https://doi.org/10.1007/978-3-030-01204-5_7

because of two reasons i.e. high-costs, and limitations in scope. Both of these reasons must intense pressure on the concerned research and the application of that research. This is the reason that scarce nature of the parallel data with respect to different languages is considered to be one of the main issues in SMT [2]. These types of corpora are not easily found, especially in the case where minimal parallel-resource language pairs are involved. Even if we analyze the cases involving the well-resource languages, such as Europarl [3], the SMT performance adopts a downward trend in significant way if it is applied to a slightly different domain. This is the reason that the efficiency of the performance decreases as the change occurs in the domain. In order to tackle with the lack of parallel data, bridge language technique, as a common solution is used. If the languages with inefficient resources are to be involved, then this issue becomes significant in relation to an SMT system. However, the most encouraging point is the sufficient availability of the resources between them and the other languages. This issue has been determined that improvement in general case does not occur as a result of intermediary languages, still this particular idea can be employed in the form of a simple method. This idea is adopted as a simple method so that translation performance for the existing systems could be enriched [4]. If we are indulged in a scenario where we have to deal with the parallel corpus between the source and target languages, we must try to improve the overall translation quality and coverage. However, this translation quality and coverage could only be improved if the direct model based on this parallel corpus is combined with a bridge model. So, increasing the information gain is a reason to propose the direct-bridge combination method.

In this paper, a combination method of direct and bridge SMT models will be proposed to apply on Persian-Spanish under-resource language pair. The basic reason for this proposal is to prevent the relevant portions of the bridge SMT model from interfering with the direct SMT model. We show positive results for our case-study on different direct training data size, and as a positive side effect, we achieve a large reduction of bridge translation model size.

2   Related Work

Recently, some efforts have been made so that the quality and recall of the bridge-based SMT could be enhanced. During one of the experiments [5] sought help from the bridge language so that word alignment system along with the procedure for combining word alignment systems could be created. They did this experiment so that these systems could be created from multiple bridge languages. When it comes to the stage of obtaining the final translation then it is conducted through consensus decoding. The entire process of consensus decoding combines hypotheses that are gained after all the bridge language word alignments are obtained. Later on, the effect of the bridge language during the final translation system was examined by [6]. They revealed through their experimentation that if the size of training data is small in any case then the bridge language should be same as the source one but if training data is large then in that case, the bridge language looks similar to the target one. Whatever the case is, it will be preferable to use bridge language with a structure similar to source and target languages. An experiment was conducted by [7], during which the researchers focused on resolving the issue through the help of source-target translations. However, the interesting fact is that these source-target translations were not generated because the source phrase and target phrase that correspond with these translations connect to different bridge phrases. One of the basic ways through bridging idea can be demonstrated is by the large-size of the newly created bridge phrase-table. Some effort has been made so that precision on bridge language technique could be improved. According to the studies conducted by [8], it has been confirmed that transitive property between three languages does not exist. So it can easily be said that most of the translations that were produced within the final phrase-table could not be right. One of the methods has been derived from the structure of source dictionaries, while the other method has been derived from the distributional similarity. A strategy has been introduced that uses context vectors so that pruning method could be created for the purpose of removing the phrase pairs. At this point, only those phrase pairs are removed that link to each other either through weak translations of through polysemous bridge phrase [9].

Our approach is similar to Domain Adaptation methods. These methods enable us to combine the training data from various sources and build a single translation model. This single translation model is then used for the purpose of translating sentences into the new domain. Various methods have been used to explore the domain adaptation within the field. Some of these methods focus on using the Information Retrieval (IR) techniques so that sentence pairs related to the target domain from a training corpus could be retrieved [10]. Other domain adaptation methods focus on creating a distinction between the examples of general and specific domain [11]. [12] during the similar scenario, used the multiple alternative decoding paths so that various translation models could be combined. They also made sure that the weights of these translation models are set using help from the Minimum Error Rate Training (MERT) [13].

3   Bridge Language Theory

High-quality data set is not always available for training the SMT systems. One of the possible ways to solve this impasse is to using a third language as a bridge one for which there exist high-quality source-bridge and bridge-target bilingual resources. A bridge language is a natural language used as an intermediary language for translation between many different languages. The bridge language technique is an idea to generate a systematic SMT when a proper bilingual corpus is lacking or the existing ones are weak. The major drawback and concern of generated translations through bridging is the translation quality, as it is possible to produce erroneous translations by transferring errors or ambiguities from a language pair to another through the bridge language. However, when language resources in specific language pairs do not exist or are scarce, the use of bridge languages as data bridges can prove to be a convenient linguistic short-cut for offering language services or building and enhancing language resources. There are methods by which the resources of bridge language can be utilized as explained in [14], namely;

1. Sentence-level bridging or transfer (cascade) approach.
2. Phrase-level bridging or triangulation (multiplication) approach.
3. Synthetic corpus approach.

3.1   Sentence-Level Bridging

The transfer approach, first, converts the source language into bridge one by translating it with the help of source-bridge translation system. After then it converts from bridge language to target one through the bridge-target translation system. Given a source sentence, s, we can also translate it into n bridge language sentences $(b_1, b_2, b_3, \ldots, b_n)$, using a source-bridge translation system. Each of these n sentences, b_i, can then be translated into m target language sentences $(t_{i1}, t_{i2}, t_{i3}, \ldots, t_{im})$, using bridge-target translation system. Thus, in total we will have $(m.n)$ target language sentences. These sentences can then be re-scored with the help of source-bridge and bridge-target translation system scores. If we denote source-bridge system features as γ^{sb} and bridge-target system features as γ^{bt}, the best scoring translation is calculated using Eq. (1):

$$\hat{t} = \arg\max_b \sum_{k=1}^{L} \left(\lambda_k^{sb} \gamma_k^{sb}(s, b) + \lambda_k^{bt} \gamma_k^{bt}(b, t) \right) \tag{1}$$

where L is the number of features used in SMT systems, λ^{sb} and λ^{bt} are the feature weights. In this approach, for assigning the best target candidate sentence, t, to the input source sentence, s, we maximize the probability $P(t|s)$ by defining hidden variable, b, which stands for the bridge language sentences, we gain:

$$\arg\max_s P(t|s) = \arg\max_s \sum_b P(t, b|s) = \arg\max_s \sum_b P(t|b, s)P(b|s) \tag{2}$$

Assuming s and t are independent given b:

$$\arg\max_s P(t|s) \approx \arg\max_s \sum_b P(t|b)P(b|s) \tag{3}$$

In Eq. (3) summation on all b sentences is difficult, so we replace it by max-imization. Eq. (4) is an estimate of Eq. (3):

$$\arg\max_s P(t|s) \approx \arg\max_s \max_b \sum_b P(t|b)P(b|s) \tag{4}$$

Instead of searching all the space of b sentences, we can just search a subspace of it. For simplicity we limit the search space in Eq. (5). A good choice is b subspace produced by the k-best list output of the first system (source-bridge):

$$\arg\max_s P(t|s) \approx \arg\max_s \max_{b \in k-best(t)} \sum_b P(t|b)P(b|s) \qquad (5)$$

In fact each source sentence, s, of the source test set is mapped to a subspace of total b space and search is done in this subspace for the best candidate sentence, t, of the second system (bridge-target).

3.2   Phrase-Level Bridging

Concerning the phrase-level bridging approach, we directly create a source-target phrase-table from a source-bridge and a bridge-target phrase-table. In this approach phrase s in the source-bridge phrase-table is connected to b, and this phrase b is associated with phrase t in the bridge-target phrase-table. We link the phrases s and t in the new phrase-table for the sourcetarget. For scoring the pair phrases of the new phrase-table, assuming $P(b|s)$ as the score of the source-bridge phrases and $P(t|b)$ as the score of the bridge-target phrases, then the score of the new pair phrases s and t, $P(t|s)$ in source-target phrase-table is counted:

$$P(t|s) = \sum_b P(t, b|s) \qquad (6)$$

b is a hidden variable and actually stands for the phrases of bridge language;

$$P(t|s) = \sum_b P(t|b, s)P(b|s) \qquad (7)$$

Assume that s and t are independent, given b:

$$P(t|s) \approx \sum_b P(t|b)P(b|s) \qquad (8)$$

For simplicity the summation on all the b phrases is replaced by maximization, then Eq. (8) is approximated by:

$$P(t|s) \approx \max_b \sum_b P(t|b)P(b|s) \qquad (9)$$

3.3   Synthetic Corpus

This method attempts to develop a synthetic source-target corpus by translating the bridge part in the source-bridge corpus, into the target language by means of a bridge-target model, and translating the bridge part in the target-bridge corpus into the source language with a bridge-source model. Eventually, it combines the source sentences with the translated target sentences or combines the target sentences with the translated source sentences. The source-target corpora that is created using the above two methods can be blended together so a final synthetic corpus could be produced. However, it is complicated to create a high-quality translation system with a corpus compiled merely by an SMT system.

4   Direct-Bridge Combination Scenario

If we seek for the best performing approach of the bridge language technique then it is called triangulation which helps in the construction of an induced new phrase-table so that source and target languages could be linked. The biggest issue encountered during the application of this approach is that the size of the bridge phrase-table is very large [15]. In this scenario we generate a new source-target translation model which is in contrast to domain adaptation. This method contains the phrase bridging (triangulation) technique from two models. But we also use the domain adaptation approach so that relevant portions of the bridge phrase-table could easily be selected. Furthermore, we improve the translation quality by combining these portions with the direct translation model. We also explore how to merge bridge and a direct model built from a given parallel corpora into an effective combination by using the optimized direct-bridge combination method. This combination will help us in enhancing the coverage and bringing an improvement to the translation quality. We take the information that is gained through the relevant portions of the bridge model and then try to maximize it. The used information do not interfere with the trusted direct model. We further ponder over the notion of categorizing the bridge phrase pairs. Later on, we divide these bridge phrase pairs into five different categories in accordance with their relation to the existence of source or target phrases in the direct model. The phrase pairs included in the first category, *cat-1*, present a combination of the source and target phrases in the direct system. The second category, *cat-2* is a bit different from the first category. The only similarity between both of the categories is that both of them contain the source and target phrases. However, the source and target phrases in the second category are not merged as a phrase pair in the direct system. The third, *cat-3*, fourth, *cat-4*, and fifth, *cat-5* categories represent the presence of source and target phrase only but none of them are involved in the direct system.

Different categories demonstrated within the Table 1 show portions that have been derived from the bridge phrase-table. These categories have been included in the Table 1 with their labels which will help us with our results.

Table 1. Phrase pairs categorization of the portions extracted from the bridge phrase-table.

Bridge phrase pairs cat	Src in direct	Trg in direct	Src and Trg in direct
cat-1	Yes	Yes	Yes
cat-2	Yes	Yes	No
cat-3	Yes	No	No
cat-4	No	Yes	No
cat-5	No	No	No

5   Experimental Framework

In this work we used the *Moses* as a phrase-based SMT decoder [16]. The combination scenario is used for creating a link between the direct model and the different bridge portions. Later on, we used an in-domain parallel corpus containing (200K) sentences (approximately (5M) words) that were derived from Open-Subtitles parallel corpus [17] for the purpose of following the direct Persian-Spanish SMT model. We also constructed two SMT models while conducting the bridge-based experiments. One model is used to create a translation from Persian language to English one, while the other model focused on translating from English to Spanish language. The English-Spanish parallel corpus contains almost (2M) sentences (approximately (50M) words) that have been derived from the Europarl corpus [3]. We use an in-domain Persian-English parallel corpus that contained almost (165K) sentences (approximately (4M) words) derived from TEP parallel corpus [18]. We used *fast-align* tool-kit for the purpose of conducting the word alignment. In the case of Spanish language modelling, almost (200M) words were derived and used from the Europarl corpus, in combination with the Spanish side of our training data. We sought help from the *Ken* language modelling [19] so that all the implemented language models could be inserted with 4-grams. In order to cater with the English language modelling, we sought help from the English side of the Europarl corpus with 4-gram LM through the Ken tool-kit as well. Moses phrase-based SMT system was specifically used for the purpose of conducting all these experiments. We also sought help from MERT when we are about to decode the weights optimization. In the scenario, where we have to tackle with both the Persian-English and English-Spanish translation models, we optimize the weights through a set of (5K) sentences. These sentences were derived from the parallel corpus and were then randomly checked for each model. While dealing with all of the models, we take care to only use the maximum phrase length of size (6) across all models. Afterwards, we report the results on an in-domain Persian-Spanish evaluation set. This set included almost (500) sentences and two references. We conducted the evaluation by using the *BLEU* metric [20]. The phrase-based Moses provides us with the flexibility to use the multiple translation tables in the case of direct-bridge combination method experiments. During the scenario, where translation options are collected from one particular table while other tables are used for the purpose of collecting the additional options, we use the *Couple* during the combination technique. However, the fact is that we can make our selection from the various options of combination techniques. If in any case, one translation option (identical source and target phrases) is found in multiple tables then we would create separate translation options for each occurrence. However, the score for each translation option will also be kept different.

5.1   Baseline Systems Evaluation

We compare the performance of sentence bridging (transfer) method against phrase-level bridging (triangulation) method with different filtering thresholds.

Generally, the triangulation method outperforms the transfer one even when we use a small filtering threshold of size (100). Moreover, the higher the threshold the better the performance but with a diminishing gain. We use the best performing set-up across the rest of the experiments which is filtering with a threshold of (10K). The results are presented in Table 2.

Table 2. Comparing the performance of transfer method and triangulation method with different filtering thresholds according to BLEU.

Bridge scheme	BLEU
Transfer	20.21
Triangulation (filtering 100)	20.64
Triangulation (filtering 1,000)	21.18
Triangulation (filtering 10,000)	21.57

According to Table 2, the triangulation method by the filtering of (10K) sentences outperforms the rest of the filtered types, and the transfer method.

5.2   Baseline Combination Systems Evaluation

We start by the basic combination approach and then explore the gain/loss achieved from dividing the bridge phrase-table to five different categories (*cat-1* to *cat-5*). Table 3 illustrates the results of the basic combination in comparison to the best bridge translation system (triangulation by (10K) filtering) and the best direct translation system.

Table 3. Baseline systems combination experiments between the best bridge-based baseline translation system and the best direct-based translation system according to BLEU.

Translation systems	BLEU
Direct	22.45
Triangulation (filtering 10,000)	21.57
Direct+Triangulation (filtering 1,000)	22.81

As an interesting observation from Table 3, direct translation system has better performance than triangulation by filtering (10K) sentences. The reason is related to the large size of parallel corpus for training the direct-based translation system. In comparison with the previous set of experiments we can see that the difference between training data sizes have a direct effect on the performance of direct translation systems. The results show that combining both models leads to a gain in performance. Now, the problem is finding a possibility to improve the quality by doing a smart choice of only relevant portion of the bridge phrase-table. We can overcome this problem through our proposed direct-bridge combination method.

5.3   Direct-Bridge Combination System Evaluation

In this portion, we will ponder over the idea of creating a division of the bridge phrase pairs into five different categories. This division will be done according to the existence of source or target phrases within the direct-based system. We first conduct a discussion of the results, and then reveal the trade-off that occurs between the quality of translation and the size of the different categories. These categories have been derived from the bridge phrase-table. Table 4 reveals the results of the direct-bridge combination method experiments that have been demonstrated on the learning curve of (100%) (approximately (200K) sentences), (25%) (approximately (50K) sentences) and (6.25%) (approximately (12.5K) sentences) of the Open-Subtitles Persian-Spanish parallel corpus.

Table 4. Optimized direct-bridge combination experiments results according to BLEU.

Translation models	12.5K sentences	50K sentences	200K sentences
Direct	15.85	20.01	22.45
Triangulation	19.89	20.18	21.57
Baseline combination	**21.72**	**22.09**	**22.81**
cat-1	**17.38**	20.20	21.96
cat-2	**18.53**	20.58	22.06
cat-3	**17.54**	20.19	**22.76**
cat-4	**18.32**	**20.93**	**23.14**
cat-5	**19.97**	**21.64**	22.45

In Table 4 the first rows are revealing the outcome of the direct-based system. The second row reveals that outcome that we have gained from the best bridge-based system (triangulation). The third row reveals the outcome of the baseline combination experiments conducted along with the pattern of whole bridge phrase-table. Furthermore, the next set of rows reveals the results of our direct-bridge combination method experiments that have been derived on the basis of a different categorization. All scores are highlighted in BLEU. The bold scores have been used to mark a statistically significant result against the direct baseline translation system.

6   Discussion

The results further reveal that bridging is basically a technique considered to be robust because no or small amount of parallel corpora is present in it. When the direct-based translation system and the bridge-based translation system merge with each other in order to form a base combination, they end-up giving a boost to the translation quality across the learning curve. So it can simply be expected that we will gain more from this combination if we use the smallest form of parallel corpus. The results also reveal that some of the bridge categories provide more information gain in comparison to the other categories. It also happens sometimes that some of the categories damage the entire quality. For instance, *cat-1* and *cat-2* both heavily contribute towards damaging the quality of translation if they are combined with direct model that has gained training on (100%) of the parallel data (approximately (200K) sentences). We have also gained an interesting observation from the results and that is we can achieve a better performance in comparison to a model trained on four times the amount of data approximately (50K) sentences if we construct a translation system with only (6.25%) of the parallel data (approximately (12.5K) sentences). Another most important point that we derive from the learning curve is that if the source phrase in the bridge phrase-table does not exist in the direct model then we can easily achieve the best gains. Such an expectation arises in the scenario where by conducting an addition of the unknown source phrases, we succeed in decreasing the overall OOVs. Creating a reduction in the bridge phrase-table is considered to be an additional benefit when we relate it with the proposed direct-bridge combination method. If we analyze the Table 5 then we will come to know that the percentage of phrase pairs is basically derived from the original bridge phrase-table so that each bridge category across the learning curve could properly be denoted.

Table 5. Percentage of phrase pairs extracted from the original bridge phrase-table for each bridge-based category.

Translation models	12.5K sentences	50K sentences	200K sentences
cat-1	0.1%	0.1%	0.2%
cat-2	16%	19%	35.2%
cat-3	64.1%	63.3%	59.9%
cat-4	6.1%	3.4%	2.3%
cat-5	13.7%	4.3%	2.3%

At this point, the group of the phrase pairs is extracted in the form of categories. This is done in order to make it clear that source phrases exist in the direct model which makes the least contribution. These source phrases also damage the overall combination performance sometimes. The direct-bridge combination scenario with target-only category provides comparatively better results in BLEU while hugely reducing the size of the bridge phrase-table used ((2.3%) of the original bridge phrase-table), if it is viewed in accordance with large parallel data (approximately (200K) sentences). However, in the case of smaller parallel data, the advantage is comparatively decreased but two new tools are introduced including the trade-off between the quality of the translation and the size of the model. We can easily create an improvement in translation quality of minimal parallel-resource SMT systems if the optimized direct-bridge combination method between bridge and direct systems are proposed. We revealed that this scenario can result in creating a large reduction of the bridge-based system without affecting the performance in any positive way.

7   Conclusion and Future Work

We applied the direct-bridge combination approach between bridge and direct models to improve the translation quality. We showed that the selective combination can lead to a large reduction of the bridge model without affecting the performance if not improving it. In the future, we plan to investigate classifying the bridge model based on morphological patterns extracted from the direct model instead of just the exact surface form.

Acknowledgment. The authors would like to express their sincere gratitude to Dr. Mojtaba Sabbagh-Jafari for his helpful comments.

References

1. Ahmadnia, B., Serrano, J., Haffari, G.: Persian-Spanish low-resource statistical machine translation through English as pivot language. In: Proceedings of the International Conference on Recent Advances in Natural Language Processing (RANLP), pp. 24–30 (2017)
2. Babych, B., Hartley, A., Sharoff, S., Mudraya, O.: Assisting translators in indirect lexical transfer. In: Proceedings of the 45th Annual Meeting of the Association of Computational Linguistics (ACL) (2007)
3. Koehn, P.: Europarl: a parallel corpus for statistical machine translation. In Proceedings of the 10th Machine Translation Summit (AAMT), Phuket, Thailand, pp. 79–86 (2005)
4. Matusov, E., et al.: System combination for machine translation of spoken and written language, pp. 1222–1237. In: Proceedings of Transactions on Audio, Speech and Language (IEEE) (2008)
5. Kumar, S., Och, F., Macherey, W.: Improving word alignment with bridge languages. In: Proceedings of the Joint Conference on Empirical Methods in Natural Language Processing (EMNLP), and Computational Natural Language Learning, pp. 42–50 (2007)
6. Paul, M., Yamamoto, H., Sumita, E., Nakamura, S.: On the importance of pivot language selection for statistical machine translation. In: Proceedings of Human Language Technologies: The Annual Conference of the North American Chapters of the Association for Computational Linguistics (HLT-NAACL), pp. 221–224 (2009)
7. Zhu, X., He, Z., Wu, H., Zhu, C.: Improving pivot-based statistical machine translation by pivoting the co-occurrence count of phrase pairs. In Proceedings of the Conference on Empirical Methods in Natural Language Processing (EMNLP), pp. 1665–1645 (2014)

8. Saralegi, X., Manterola, I., Vicente, I.: Analyzing methods for improving precision of pivot based bilingual dictionaries. In: Proceedings of the Conference on Empirical Methods in Natural Language Processing (EMNLP), pp. 846–856 (2011)

9. Tofighi, S., Bakhshaei, S., Khadivi, S.: Using context vectors in improving a machine translation system with bridge language. In: Proceedings of the 51st Annual Meeting of the Association for Computational Linguistics (ACL), pp. 318–322 (2013)

10. Hildebrand, A., Eck, M., Vogel, S., Waibel, A.: Adaptation of the translation model for statistical machine translation based on information retrieval. In Proceedings of the Conference on Empirical Methods in Natural Language Processing (EMNLP) (2005)

11. Daume, H., Marcu, D.: Domain adaptation for statistical classifiers. J. Artif. Intell. (JAIR) **26**, 101–126 (2006)

12. Koehn, P., Schroeder, J.: Experiments in domain adaptation for statistical machine translation. In: Proceedings of Workshop on Machine Translation (WMT) (2007)

13. Och, F.: Minimum error rate training in statistical machine translation. In: Proceedings of the 41st Annual Meeting of the Association for Computational Linguistics (ACL) (2003)

14. Wu, H., Wang, H.: Pivot language approach for phrase-based statistical machine translation. In: Proceedings of the 45th Annual Meeting of the Association for Computational Linguistics (ACL), pp. 856–863 (2007)

15. Elkholy, A., Habash, N., Leusch, G., Matusov, E.: Selective combination of pivot and direct statistical machine translation models. In: Proceedings of the International Joint Conference on Natural Language Processing, Nagoya, Japan, pp. 1174–1180 (2013)

16. Koehn, P., et al.: Moses: open source toolkit for statistical machine translation. In: Proceedings of the 45th Annual Meeting of the Association for Computer Linguistics (ACL), pp. 177–180 (2007)

17. Tiedemann, J.: Parallel data, tools and interfaces in OPUS. In: Proceedings of the 8th International Conference on Language Resources and Evaluation (LREC) (2012)

18. Pilevar, M., Faili, H., Pilevar, A.: TEP: Tehran English-Persian parallel corpus. In: Proceedings of 12th International Conference on Intelligent Text Processing and Computational Linguistics (CICLing) (2011)

19. Heafield, K.: KenLM: faster and smaller language model queries. In: Proceedings of the 6th Workshop on Statistical Machine Translation, pp. 187–197 (2011)

20. Papineni, K., Roukos, S., Ward, T., Zhu, W.: BLEU: a method for automatic evaluation of machine translation. In: Proceedings of the 40th Annual Meeting on Association for Computational Linguistics (ACL), pp. 311–318 (2002)

Automatic Mining of Discourse Connectives for Russian

Svetlana Toldova[1], Maria Kobozeva[2], and Dina Pisarevskaya[2]([✉])

[1] National Research University "Higher School of Economics", Moscow, Russia
stoldova@hse.ru
[2] Institute for Systems Analysis FRC CSC RAS, Moscow, Russia
kobozeva@isa.ru, dinabpr@gmail.com

Abstract. The identification of discourse connectives plays an important role in many discourse processing approaches. Among them there are functional words usually enumerated in grammars (*iz-za* 'due to', *blagodarya* 'thanks to',) and not grammaticalized expressions (*X vedet k Y* 'X leads to Y', *prichina etogo* 'the cause is'). Both types of connectives signal certain relations between discourse units. However, there are no ready-made lists of the second type of connectives. We suggest a method for expanding a seed list of connectives based on their vector representations by candidates for not grammaticalized connectives for Russian. Firstly, we compile a list of patterns for this type of connectives. These patterns are based on the following heuristics: the connectives are often used with anaphoric expressions substituting discourse units (thus, some patterns include special anaphoric elements); the connectives more frequently occur at the sentence beginning or after a comma. Secondly, we build multi-word tokens that are based on these patterns. Thirdly, we build vector representations for the multi-word tokens that match these patterns. Our experiments based on distributional semantics give quite reasonable list of the candidates for connectives.

Keywords: Rhetorical Structure Theory · Discourse connectives
Word embeddings

1 Introduction

The automatic detection and extraction of discourse relations is one of the essential tasks of NLP. It can significantly improve the performance of several Natural Language Processing applications, e.g. deception and intent detection [15,17], summarization [11,16], sentiment analysis [10,13,23], question-answering [7,25], argumentative discourse analysis [8] and etc., besides, it is widely used for building text generation systems. One of the approaches to the task is to use a list of special cues that signal certain types of discourse relations. There are closed classes of lexemes and multi-word expressions (functional words), such as

The study was funded by RFBR according to the research project 17-29-07033.

conjunctions, prepositions and others that mark certain types of rhetoric relations (e.g. the conjunction *because* or the preposition *for* can signal the 'cause-effect' relation between discourse units). Though the lists of these expressions are presented in grammars and some other sources for Russian, there are no ready-made lists of connectives mapped onto certain types of discourse relations. The problem is that only lexicalized items are usually mentioned in dictionaries and grammars. Such connectives usually signal intra-sentential relations between clauses, and not between larger discourse units. As various studies of discourse has shown, there are less grammaticalized multi-word expressions denoting relations between discourse units (e.g. expressions containing content words, such as *prichina etogo* 'the cause is' or *eto ob'yasnyaet to, chto* 'this explains, that') [6,18]. Our research is devoted to this type of connectives. The task is to compile a list of markers, or to be more precise, to enhance a seed list.

The most up-to-date approach to expanding a list of lexemes with semantically close words is to use word embeddings for this task. One can take open pre-trained Word2Vec models to compile such lists (cf. models from http://rusvectores.org/ru/models/). As for connectives under discussion, there are several obstacles to use this method. Firstly, functional words usually have no independent meaning, they are usually in stop-lists and, thus, are absent in the models. Secondly, even so-to-called primary markers are often multi-word expressions and not unigrams. Thirdly, the usage of the models is based on the 'distributional hypothesis' [9] that semantically similar words occur in similar contexts. However, even the connectives containing content words express the structural relations between text spans, rather than denote some concepts. As a result, their contexts are semantically heterogeneous.

In our work we conduct series of experiments to examine the applicability of Word2Vec models to the task of finding multi-word connectives signaling the same discourse relation as seed functional words from a dictionary. We suggest some linguistically motivated heuristics which can help to overcome the mentioned above issues. Our method is based on the assumptions that (1) discourse units can be substituted by some special anaphoric expressions in the context of the majority of discourse markers (e.g. *po prichine etogo* 'because of this'); (2) the discourse markers have a tendency to occur at the beginning of a sentence, or after a comma, some of the constructions occur before a comma; (3) there are several pattens based on content words denoting certain discourse relations that were singled out in previous works on the basis of Russian Rhetorical Structure Treebank (Ru-RSTreebank). We glue multi-word conjunctions into one token. We also glue n-grams (3-grams) situated in the mentioned above positions and containing special anaphoric expressions into corresponding tokens. And after this, we train Word2Vec models. As our experiments has shown, this method gives a satisfactory result, though it needs further elaboration.

We focus only on one type of discourse relations, namely, 'cause-effect' relations. However, the suggested technique can be used for mining other kinds of discourse markers.

The paper is organized as follows. Following a discussion of background of our research in Sect. 2, we describe our methods in Sect. 3. In this section we introduce our dataset, describe the necessary steps of preprocessing the data and also the models we have used. In Sect. 4, we present the results of our experiments and finally we conclude the paper in Sect. 5.

2 Background

2.1 Rhetorical Structure Theory Approach to Discourse Connectives

Our study is based on the discourse representation within the framework of the Rhetorical Structure Theory [12,22]: discourse is organized as a hierarchical system of discourse units of different size, where smaller discourse units can be embedded into larger ones, in case there is a rhetorical (discourse) relation of a certain type between them, e.g. 'concession', 'cause-effect', 'elaboration' etc. The exemplified relations are asymmetric ones (between nucleus and satellite). Elementary discourse units (EDUs) usually correspond to clauses. Consequently, the 'cause-effect' relation between discourse units correspond to causal relations between facts. Facts can be expressed via phrases headed by non-finite verb-forms such as infinitives or nominalizations (c.f. *his singing yesterday...*). Some types of phrases (smaller than a finite clause) are also treated as EDUs. This approach is widely spread in NLP systems and in RST Treebanks annotation rules for written texts (cf. [4,5,20,24] and others).

2.2 Discourse Connectives Features

As it has been mentioned, there are special clues in discourse signalling that there is a relation between two discourse units. Some of them are functional words (linking words), that are lexemes whose primary function is to express different relationships between pieces of texts - for example, *vsledstviye* 'in consequence of'. In [19] such clues are considered as primary connectives. Less grammati-calized, secondary connectives are, mostly, multi-word expressions, such as *eto privelo k tomu, chto* 'this led to the fact that'. Secondary connectives are currently under-represented in lexicons. As discourse connectives are reliable signals for different types of relations, building a lexicon of the connectives, focusing on secondary connectives, is an essential task.

Salient locations for discourse connectives in textual structure are beginning of paragraph, beginning of the sentence, a position after a punctuation mark or immediately before it [1].

In Russian, some primary connectives are listed in Russian grammar [21]. Content words expressing cause-effect relation are presented in [2,3], in Rus-gram [14]. A list of less-grammaticalized connectives is discussed in [24]. These connectives were extracted manually from the Ru-RSTreebank corpus. The basic patterns for multi-word connectives formation were singled out.

Ru-RSTreebank (http://linghub.ru/ru-rstreebank/) is annotated for discourse relations. Its first part, that we used for the current research, consists of 79 texts: news stories, news analytics and popular science (5582 EDUs and 49840 tokens in total). There are 330 examples of causal relations there: 220 examples for the Cause-Effect relation and 110 for the Evidence.

The aim is to expand the lexicon via mining new items (not listed in grammars and dictionaries) from a large corpus on the basis of a short seed set of connectives from dictionaries and of a bigger set of connectives extracted manually from the Ru-RSTreebank.

2.3   Event Anaphora

There are anaphoric elements in Russian that refer to an event and not to an entity (hence fore, substituting a whole discourse unit) such as *eto* 'this, neutrum' and *chto* 'what'. There are also so-to-called correlative expressions (expressions that serve as connectives in relative clauses) such as *to, chto* 'the fact, that'. They are treated as parts of some grammaticalized connectives, or they substitute connectives arguments. Hence, the connectives are two-argument predicates; arguments are discourse units (facts); arguments can be replaced with anaphoric elements:

[V SSHA v 80-kh gg. proizoshel perekhod s primeneniya aspirina na paratsetamol sredi detey.] [Eto moglo stat' prichinoy uvelicheniya chisla detey, zabolevshikh astmoy v techeniye dannogo perioda.] [In the US in the 80's there was a transition from aspirin to paracetamol among children.] [This could have caused an increase in the number of children who developed asthma during this period.]

The anaphoric expressions are quite frequent in the corpus. Thus, they can be taken into consideration in connectives mining.

2.4   Patterns for Connectives Lexicon Construction

Some of the multi-word connectives have a content word denoting a particular rhetoric relation as a core word (e.g. *prichina* 'cause' in *po prichine*). They can take one of the mentioned above expression (anaphoric or correlative) as an argument.

The following basic patterns can be used: 1. n-gram + *to, chto* 'that': *po prichine togo, chto* 'by reason that'. 2. n-gram: includes *eto* 'this': *po prichine etogo* 'by this reason', *v resultate etogo* 'as a result of this', *privodit k tomu, chto* 'lead to the fact that'. 3. n-gram + *chto* 'what': *v resultate chego* 'lit. as the result of what', *chto yavlyayetsya prichinoy* 'that is the cause'.

Basic patterns are connected with classes of core word: 1. prepositions: *iz-za togo, chto* 'by reason of'; 2. adverbial phrases (preposition + content noun): *po prichine togo, chto* 'due to the fact that'; 3. causative verbs (verbs of causation, mental impact, change of state) + (preposition): *X vedet k Y* 'X leads to Y', *X vlechet Y* 'X enatails Y', *X govorit o tom, chto* 'X says that'; 4. light verb constructions (lexical functions): *X privelo k rezul'tatu* 'X led to the result', *X yavlyayetsya rezul'tatom Y* 'X accounts for', lit. 'to be a cause'.

We can test different methods for extracting connectives using these patterns.

3   Dataset and Method

3.1   Initial List of Connectives

Two seed lists of causal relation connectives are available. The first, initial one is taken from lexicons and contains 11 elements: *blagodarya* 'thanks to', *v rezul'tate* 'as a result of', *v svyazi s* 'in connection with', *vvidu* 'due to', *vsledstviye* 'in consequence of', *iz-za* 'due to', *poskol'ku* 'since', *potomu chto* 'because', *tak kak* 'as', *tak chto* 'so that', *poetomu* 'therefore'. The bigger seed list from the Ru-RSTreebank can be expanded by multi-word expressions, such as 'a verb + anaphoric expression', e.g. *eto prived'ot k* 'this will lead to', *iz chego sleduet* 'lit. from what follows', etc. [24].

3.2   Dataset Initial Preprocessing and Models

The dataset for the experiments contains news texts in Russian (approx. 285 mln tokens). Texts were not lemmatized during the preprocessing steps: we used definite word forms in our search patterns. In the existing Word2Vec models, some discourse connectives are considered as stop-words and removed when the model is trained on context words - for example, conjunctions *potomu* and *potomu chto* 'because' are not included in http://rusvectores.org/ru/ models, thus, they cannot be used for the task of discourse connectives mining. For connectives list expansion we train word2vec's 'skip-gram model' using GenSim. Punctuation marks and almost all stop-words are not removed from the texts before model training, because they could signal that the EDU is connected with the previous one - for example, as in anaphoric expression *blagodarya etomu* 'due to this' - and could be included as parts in discourse markers. The only removed stop words are particles and some very frequent conjunctions *zhe, zh, nu* 'well', *a* 'but', *dazhe* 'even', *lish* 'only', *i* 'and', that are frequently placed between words in the connectives.

We run two different experiments, based on two different Word2Vec models.

1. In the first experiment we use all the connectives from the small seed list (11 connectives). For multi-word connectives, such as *potomu chto* 'because', we glue each one of them into one token, in the whole dataset, before the model training. It is the single change in the dataset, after initial preprocessing. Thus, the model was trained on the multi-word tokens for seed set and unigrams for the rest of the corpus. After that we search for connectives, that are semantically similar to the small seed list connectives (we consider top 20 connectives for each seed connective).

2. For the second experiment, in addition to concatenating multi-word connectives from the small seed list into one token, we also make further dataset preprocessing: we also concatenate all 3-grams situated in the positions that correspond to the patterns and contain special anaphoric expressions, into single tokens. Thus, we have multi-word tokens for the seed and for the patterns.

The following patterns are used for multi-word constructions concatenation in the dataset: a. the construction of 3-grams is in the beginning of the sentence and begins with possible word forms of *eto* 'this'; b. the construction of 3-grams follows after a comma and begins with possible forms of *chto* 'what'; c. 3-grams that contain forms of *to* 'that'. After them there is a comma, after the comma there are *chto, kak*.

After that, we search for connectives that are semantically similar to the initial ones.

Three patterns reflect three types of discourse units 'anaphoric' substitution: correlative *to, chto* 'the fact, that' type; sentence initial *eto* 'this' type; clause initial intra-sentence type *chto* 'what'.

According to our hypothesis, we assume that new connectives from the extended list should also be multi-word, such as *vsledstviye togo, chto* 'in consequence of', *na osnovanii togo, chto* 'on the basis of', *po prichine togo, chto* 'for the reason that'. So the results for 'concatenated' tokens in the second experiment should be better than the results in the first experiment, as they let find such expressions.

4 Experiment

We get the lists of most similar expressions for each connective from the seed. The expressions were assessed by an expert whether they can be considered as cause-effect connectives. The precision was calculated for each connective as a proportion of the true positive expressions in the list of 20 most similar ones. Then, the average precision was calculated. For the two experiments, we get the following results (Table 1).

Table 1. Results for Word2Vec models.

Model	Precision
1. Multi-word tokens for seed set	38%
2. Multi-word tokens for seed set and for patterns	54%

To compare the results, we also created frequency lists for 'to, chto', 'eto' and 'chto', based on the patterns for trigrams. It is used as a simple baseline. In the top of such frequency lists, we can also see real discourse connectives.

But such lists need further processing and show worse results before it, than the models in the current research (not more than 30% of the output are real discourse connectives).

As for the first experiment, only the three lists have the precision more, than 50%. For the second experiment, there are six lists for which the precision exceeds 50%. As a result, some primary connectives are added to the seed list such as *ibo* 'as' or *ved'* 'indeed'. The second model expands the lists via suggesting light verb expressions, cf. *v svyazi s* 'in connections with' in the seed list - *eto proizoshlo v svyazi s* 'this has happened in the connection with' *eto bylo svyazano s* 'this was due to the fact'.

5   Conclusions and Discussion

Our experiments support the initial assumptions concerning connectives features, namely, their place in sentences and their lexical properties as well as the assumption concerning the anaphoric elements that can substitute EDUs. The word2vec model that is based on the multi-word tokens including these elements gives quite reasonable list of the candidates for connectives, on the example of 'cause-effect' relations connectives.

We are planning to make further steps to understand whether distributional semantic methods and anaphoric elements for discourse units can help in the compilation of lists of patterns for multi-word connectives. Hence, we will continue with building the same model as in the second experiment, but for the bigger connectives seed list.

In future studies, we also plan to use the proposed approach, definite patterns and word embeddings models to expand seed lists of connectives for other types of rhetorical relations.

References

1. Alonso, L., Castellón, I., Gibert, K., Padró, L.: An empirical approach to discourse markers by clustering. In: Escrig, M.T., Toledo, F., Golobardes, E. (eds.) CCIA 2002. LNCS (LNAI), vol. 2504, pp. 173–183. Springer, Heidelberg (2002). https://doi.org/10.1007/3-540-36079-4_15
2. Apresyan, Y.D.: System-forming meanings to know and to consider in russian /sistemoobrazuyushchiye smysly znat' i schitat' v russkom yazyke. In: Russian Language and Linguistic Theory /Russkiy yazyk v nauchnom osveshchenii, vol. 1, pp. 5–26 (2001)
3. Boguslavskaya, O.Y., Levontina, I.B.: Meanings cause and purpose in natural language /smysly 'prichina' i 'tsel' v yestestvennom yazyke. In: Topics in the study of language /Voprosy yazykoznaniya, vol. 2, pp. 68–88 (2004)
4. Carlson, L., Marcu, D.: Discourse tagging reference manual. Technical report, ISI-TR-545, University of Southern California Information Sciences Institute (2001). http://www.isi.edu/~marcu/discourse/tagging-ref-manual.pdf

5. Carlson, L., Marcu, D., Okurowski, M.E.: Building a discourse-tagged corpus in the framework of rhetorical structure theory. In: Proceedings of the Second SIGdial Workshop on Discourse and Dialogue, SIGDIAL 2001, vol. 16, pp. 1–10. Association for Computational Linguistics, Stroudsburg (2001). https://doi.org/10.3115/1118078.1118083

6. Crible, L.: Discourse markers and (dis) fluency across registers: a contrastive usage-based study in English and French. Ph.D. thesis, UCL-Université Catholique de Louvain (2017)

7. Ferrucci, D., et al.: Building watson: an overview of the DeepQA project. AI Mag. **31**(3), 59–79 (2010)

8. Galitsky, B., Ilvovsky, D., Kuznetsov, S.O.: Detecting logical argumentation in text via communicative discourse tree. J. Exp. Theor. Artif. Intell. **30**, 1–27 (2018)

9. Harris, Z.S.: Distributional structure. In: Harris, Z.S. (ed.) Papers in Structural and Transformational Linguistics, pp. 775–794. Springer, Dordrecht (1970). https://doi.org/10.1007/978-94-017-6059-1

10. Heerschop, B., Goossen, F., Hogenboom, A., Frasincar, F., Kaymak, U., de Jong, F.: Polarity analysis of texts using discourse structure. In: Proceedings of the 20th ACM International Conference on Information and Knowledge Management, pp. 1061–1070. ACM (2011)

11. Louis, A., Joshi, A., Nenkova, A.: Discourse indicators for content selection in summarization. In: Proceedings of the 11th Annual Meeting of the Special Interest Group on Discourse and Dialogue, pp. 147–156. Association for Computational Linguistics (2010)

12. Mann, W.C., Thompson, S.A.: Rhetorical Structure Theory: Description and Construction of Text Structures. In: Kempen, G. (ed.) Natural Language Generation, pp. 85–95. Springer, Dordrecht (1987). https://doi.org/10.1007/978-94-009-3645-4_7

13. Mukherjee, S., Bhattacharyya, P.: Sentiment analysis in Twitter with lightweight discourse analysis. In: Proceedings of COLING 2012, pp. 1847–1864 (2012)

14. Pekelis, O.Y.: Causal subordinate clauses /prichinnyye pridatochnyye. In: Materials for the Project of Russian Grammar Corpus Description /Materialy dlya proyekta korpusnogo opisaniya russkoy grammatiki (2014). http://rusgram.ru

15. Pisarevskaya, D.: Rhetorical structure theory as a feature for deception detection in news reports in the Russian language. In: Computational Linguistics and Intellectual Technologies, pp. 184–193 (2017)

16. Ribaldo, R., Akabane, A.T., Rino, L.H.M., Pardo, T.A.S.: Graph-based methods for multi-document summarization: exploring relationship maps, complex networks and discourse information. In: Caseli, H., Villavicencio, A., Teixeira, A., Perdigão, F. (eds.) PROPOR 2012. LNCS (LNAI), vol. 7243, pp. 260–271. Springer, Heidelberg (2012). https://doi.org/10.1007/978-3-642-28885-2_30

17. Rubin, V.L., Conroy, N.J., Chen, Y.: Towards news verification: deception detection methods for news discourse. In: HICSS 2015 (2015)

18. Rysová, K., Rysová, M.: Discourse connectives and reference. In: TextLink2018-Final Action Conference, p. 122 (2018)

19. Rysova, M., Mírovský, J.: Use of coreference in automatic searching for multiword discourse markers in the Prague dependency treebank. In: LAW VIII - The 8th Linguistic Annotation Workshop, pp. 11–19 (2014)

20. Schauer, H.: From elementary discourse units to complex ones. In: Proceedings of the 1st SIGdial Workshop on Discourse and Dialogue, vol. 10, pp. 46–55. Association for Computational Linguistics (2000). https://doi.org/10.3115/1117736.1117742. http://portal.acm.org/citation.cfm?doid=1117736.1117742

21. Shvedova, N.Y. (ed.): Russian Grammar [Russkaya grammatika]. Nauka, Moscow (1980)
22. Taboada, M., Mann, W.C.: Applications of rhetorical structure theory. Discourse Stud. **8**(4), 567–588 (2006). https://doi.org/10.1177/1461445606064836
23. Taboada, M., Voll, K., Brooke, J.: Extracting sentiment as a function of discourse structure and topicality (2008)
24. Toldova, S., Pisarevskaya, D., Kobozeva, M.: The cues for rhetorical relations in Russian: cause-effect relation in Russian rhetorical structure treebank. Comput. Linguist. Intellect. Technol. **17**(24), 748–761 (2018)
25. Verberne, S., Boves, L., Oostdijk, N., Coppen, P.A.: Evaluating discourse-based answer extraction for why-question answering. In: Proceedings of the 30th Annual International ACM SIGIR Conference on Research and Development in Information Retrieval, pp. 735–736. ACM (2007)

Corpus Linguistics

Avoiding Echo-Responses
in a Retrieval-Based Conversation System

Denis Fedorenko$^{(\boxtimes)}$ [ID], Nikita Smetanin[ID], and Artem Rodichev[ID]

Replika.ai @ Luka, Inc., Moscow, Russia
{denis,nikita,artem}@replika.ai

Abstract. Retrieval-based conversation systems generally tend to highly rank responses that are semantically similar or even identical to the given conversation context. While the system's goal is to find the most appropriate response, rather than the most semantically similar one, this tendency results in low-quality responses. We refer to this challenge as the echoing problem. To mitigate this problem, we utilize a hard negative mining approach at the training stage. The evaluation shows that the resulting model reduces echoing and achieves better results in terms of Average Precision and Recall@N metrics, compared to the models trained without the proposed approach.

Keywords: Dialog modeling · Response selection · Lexical repetition
Hard negative mining · End-to-end learning

1   Introduction

The task of a retrieval-based conversation system is to select the most appropriate response from a set of responses given the input context of a conversation. The context is typically an utterance or a sequence of utterances produced by a human or by the system itself. Most of the state-of-the-art approaches to retrieval-based conversation systems are based on deep neural networks (NNs) [14,16]. Under these approaches, the typical response selection pipeline consists of the following steps [2]:

1. Encode the given context and pre-defined response candidates into numeric vectors, or thought vectors, using NNs;
2. Compute the value of a matching function (matching score) for each pair consisting of a context vector and each response candidate;
3. Select the response candidate with the highest matching score.

During step 1, in order to obtain thought vectors that fairly represent semantics of input contexts and responses, the conversation model is preliminarily trained to return high matching scores for true context-response pairs and low for false ones.

D. Ustalov et al. (Eds.): AINL 2018, CCIS 930, pp. 91–97, 2018.
https://doi.org/10.1007/978-3-030-01204-5_9

The challenge we faced while building the above pipeline was that the resulting model often returned high matching scores for semantically similar contexts and responses. Consequently, the model frequently repeated or rephrased input contexts instead of giving quality responses.

Consider the following conversations:

A. Context: "What is the purpose of living?"
   Response: "What is the purpose of existence?"
B. Context: "What is the purpose of living?"
   Response: "It's a very philosophical question."

The effect of rephrasing, or echoing, in conversation A in contrast to the appropriate response in conversation B can be explained by the above pipeline. It is a result of the fact that contexts and responses often contain the same concepts [4,13], hence during training on conversational datasets the NNs simply end up trying to fit the semantics of the input. The similar effect, named "lexical repetition", was also observed in [9].

In this paper, we suggest a simple and natural solution to the echoing problem for end-to-end retrieval-based conversation systems. Our solution is based on a widely used hard negative mining approach [10], which forces the conversation model to produce low matching scores for similar contexts and responses.

The paper is organized as follows. First, we describe the hard negative mining method and how we utilize it to overcome the echoing problem. Then, we introduce the evaluation metrics, our results and benchmarks for the echoing problem. We also provide the evaluation dataset used in the experiments for further research.

2   Hard Negative Mining

Let $D = \{(c_i, r_i)\}$, $i \in \{1..N\}$ be a dataset of conversational context-response pairs, where c_i, r_i – i-th context and response, respectively.

Our goal is to build a conversation model $M : (context, response) \to \mathbb{R}$ that satisfies the following condition:

$$M(c_i, r_i) > M(c_i, r_j) \tag{1}$$

$\forall i, j \neq i$ and r_j is not an appropriate response for c_i. In other words, the resulting model should return a higher matching score for appropriate responses than for inappropriate ones.

To train this model, we also need false context-response pairs as negative examples in addition to the positive ones presented in D. Consider two approaches to obtain the negative pairs: random sampling and hard negative mining. Under the first approach, we randomly select r_j from D for each c_i. If D is large and diverse enough, then a randomly selected r_j is almost always inappropriate for a corresponding c_i.

In contrast to random sampling, hard negative mining imposes a special constraint on responses selected as negatives. Let M_0 be a conversation model trained on random pairs used as negative training examples. Then, we search for a new set of negative pairs (c_i, r_j), so that their matching score satisfies the following condition:

$$M_0(c_i, r_i) - M_0(c_i, r_j) \leq m \tag{2}$$

where m is a margin (hyperparameter) between the scores of positive and negative pairs [3]. The new set of pairs is used to train the next model M_1, which, in turn, used to search for negative pairs to train M_2, and so on [1].

The intuitive idea behind hard negative mining is to select only negatives that have relatively high matching scores, and thus can be interpreted as errors of the conversation model. As a result, the model converges faster compared to random sampling [10].

Following this intuition, we can solve the echoing problem by considering contexts as possible responses, therefore the pairs (c_i, c_i) can be selected as hard negatives. In the next section, we demonstrate that this approach can ultimately prevent the conversation model from assigning a high rank to responses that are similar to contexts.

3 Experiments

For our experiments, we implement a model similar to Basic QA-LSTM described in [12]. It has two bidirectional LSTMs of size 2048 (1024 units in each direction), with separate sets of weights that encode a context and a response independently. We use a max pooling operation to calculate final thought vectors of these LSTMs. We use a cosine similarity as the output matching function. We represent input words as embeddings of size 256, which are initialized by the pre-trained word2vec vectors [8] and are not updated further during the model training. Word sequences longer than 20 words are trimmed from the right, and the context encoder is fed with only one dialog step at a time.

3.1 Models

In order to study the impact of hard negative mining on the echoing problem, we train three models using the following strategies: random negative sampling (RN), hard negative mining based on responses only (HN_r), and hard negative mining based on both responses and contexts (HN_{r+c}). We also consider the following baseline approach (BL): we use RN model to rank responses in the testing stage and then just filter out responses equal to the given context.

3.2 Datasets

We train the models on 79M of tweet-reply pairs from a Twitter data archive[1].

We perform an evaluation based on our own dataset[2]. This dataset consists of 759 context-response pairs from human text conversations, where context and

---

[1] https://archive.org/details/twitterstream.
[2] https://github.com/lukalabs/replika-research/tree/master/context-free-dataset.

Table 1. Evaluation dataset sample (see Sect. 3.2)

Context	Response
What happened to your car?	I got a dent in the parking lot
The beatles are the best	They are the best musical group ever
I'm joining the army	You're kidding. You might get killed

response both consist of a single sentence (see Table 1). We split the dataset into validation and test subsets consisting of 250 and 509 pairs, respectively. We use this dataset because it is clear, diverse and covers multiple topics of real-life conversations. Also we find it suitable for validating the echoing problem, as well as for estimating the overall model quality.

3.3 Training

The models are trained with the Adam optimizer [5] with the size of mini-batches set to 512. Intermediate models that show the highest values of the Average Precision metric on the validation set (see Sect. 3.4) are selected as the resulting models.

We use a triplet loss [3] as an objective function:

$$max(0, m - M(c_i, r_i) + M(c_i, r_j)) \tag{3}$$

where the margin m is set to 0.05. For each positive pair (c_i, r_i), a negative (c_i, r_j) is only selected within the current mini-batch using an intermediate model M trained by the moment of this batch. We only select the hard negative r_j with the highest matching score $M(c_i, r_j)$ satisfying the following condition:

$$0 \leq M(c_i, r_i) - M(c_i, r_j) \leq m \tag{4}$$

The constraint $0 \leq M(c_i, r_i) - M(c_i, r_j)$ is used to filter out the "hardest" negatives, which in practice affect convergence and lead to bad local optima [10].

We noticed that while training the HN_{r+c} model, the fraction of (c_i, c_i) negative pairs constitute up to 50% of the mini-batch.

3.4 Evaluation Methodology and Metrics

For each $context_i$ from the evaluation set, we compute matching scores for all available pairs $(context_i, answer)$, where $answer$ comes not only from the responses, but also from the all available contexts. To evaluate these results, we sort the answers by the matching score in descending order and compute the following metrics: Average Precision [7], Recall@2, Recall@5, and Recall@10 [6]. The last three metrics are indicator functions that return 1, if the ground-truth response occurs in the top 2, 5 and 10 candidates, respectively. We also introduce the context echoing metrics:

Table 2. Evaluation results based on the context-response test set (See footnote 2).

	RN	BL	HN_r	HN_{r+c}
Average Precision	0.12	0.16	0.13	**0.17**
Recall@2	0.18	0.26	0.23	**0.29**
Recall@5	0.36	0.37	0.4	**0.43**
Recall@10	0.45	0.48	**0.54**	0.53
$rank_{context}$	0.9	-	0.49	**19.43**
$diff_{top}$	0.008	-	0.01	**0.07**
$diff_{response}$	-0.15	-	-0.25	**-0.09**

- $rank_{context}$ – position (starting from zero) of the input context in the sorted results. The greater the rank, the less the model tends to return the input context among the top results
- $diff_{top}$ – difference between the top result score and the input context score. The greater the difference, the less the model tends to return relatively high scores for the context
- $diff_{response}$ – difference between the ground-truth response score and the input context score. The greater the difference, the less the model tends to return similar scores for the ground-truth response and for the context

For each metric, we compute the overall quality as an average across all test contexts. Note that for BL model we don't present context echoing metrics, since echo-responses are filtered out from the results in this approach.

3.5   Results

The results of the evaluation based on the test set are presented in Table 2. As we can see, the proposed HN_{r+c} model achieves the highest values in almost all metrics compared to other approaches. According to $rank_{context}$, it turns out that this model does not tend to highly rank input contexts and have them in the top response candidates. Still, according to the $diff_{response}$ metric, the average score of a ground-truth response is lower than the score of a context, which means that the context can be ranked higher than the ground-truth response.

We also studied the model's output. Examples of top-ranked responses for different contexts are presented in Table 3. As we can see, oftentimes the RN and HN_r models select identical or very similar responses, while the proposed HN_{r+c} model selects appropriate responses that are not necessarily semantically similar to the context. Based on this observation, we suggest that the proposed model filters out not only exact copies of the context, but also candidates with similar semantics.

Table 3. Top 3 responses for a few input contexts sorted by matching score.

RN	HN_r	HN_{r+c}
Input: What is the purpose of dying?		
1. What is the purpose of dying?	1. What is the purpose of dying?	1. To have a life
2. The victim hit his head on the concrete steps and died	2. What is the purpose of living?	2. When you die and go to heaven, they will offer you beer or cigarettes
3. To have a life	3. What is the purpose of existence?	3. It is to find the answer to the question of life
Input: What are your strengths?		
1. What are your strengths?	1. What are your strengths?	1. Lust, greed, and corruption
2. Lust, greed, and corruption	2. What are your three weaknesses?	2. I'm a robot. a machine. 100% ai. no humans involved
3. A star	3. What do you think about creativity?	3. Dunno. i mean, i'm a robot, right? robots don't have a gender usually
Input: I can't wait until i graduate		
1. I can't wait until i graduate	1. I can't wait until i graduate	1. What college do you go to?
2. What college do you go to?	2. What college do you go to?	2. School is hard this year
3. School is hard this year	3. How many jobs have you had since leaving university?	3. What subjects are you taking?

4   Related Work

In the previous works on dialog systems there was not enough attention paid to the echoing problem. The possible reason for this are "soft" evaluation conditions: test samples are constructed from a relatively small number of negative responses [3,6,14] which usually do not "echo" the test context. In [9] the "lexical repetition" is regularized by utilizing a word overlap feature during training a SMT-based dialog system. In [11,13,14] the echoing is avoided by considering only responses the dataset's contexts of which have high TF-IDF similarity with the given context. However, the latter approach is not applicable if only a set of responses is available for ranking during the testing stage, which can be the case for some domains and applications [15].

5   Conclusion

In this study, we applied a hard negative mining approach to train a retrieval-based conversation system to find a solution to the echoing problem, that is, to reduce inappropriate responses that are identical or too similar to the input context. In addition to responses, we consider contexts themselves as possible hard negative candidates. The evaluation shows that the resulting model avoids echoing the input context, tends to select candidates that are more appropriate as responses and achieves better results in terms of Average Precision and Recall@N metrics compared to the models trained without the proposed approach.

References

1. Canévet, O., Fleuret, F.: Efficient sample mining for object detection. In: Proceedings of the 6th Asian Conference on Machine Learning (ACML), No. EPFL-CONF-203847 (2014)
2. Chen, H., Liu, X., Yin, D., Tang, J.: A survey on dialogue systems: recent advances and new frontiers. SIGKDD Explor. Newsl. **19**(2), 25–35 (2017). https://doi.org/10.1145/3166054.3166058
3. Feng, M., Xiang, B., Glass, M.R., Wang, L., Zhou, B.: Applying deep learning to answer selection: a study and an open task. In: 2015 IEEE Workshop on Automatic Speech Recognition and Understanding (ASRU), pp. 813–820. IEEE (2015)
4. Jurafsky, D., Martin, J.: Dialog systems and chatbots. In: Speech and Language Processing, vol. 3 (2017). https://web.stanford.edu/~jurafsky/slp3/29.pdf
5. Kingma, D., Ba, J.: Adam: a method for stochastic optimization. arXiv preprint arXiv:1412.6980 (2014)
6. Lowe, R., Pow, N., Serban, I., Pineau, J.: The ubuntu dialogue corpus: a large dataset for research in unstructured multi-turn dialogue systems. CoRR abs/1506.08909 (2015). http://arxiv.org/abs/1506.08909
7. Manning, C.D., Raghavan, P., Schütze, H.: Introduction to Information Retrieval. Cambridge University Press, New York (2008)
8. Mikolov, T., Chen, K., Corrado, G., Dean, J.: Efficient estimation of word representations in vector space. CoRR abs/1301.3781 (2013). http://arxiv.org/abs/1301.3781
9. Ritter, A., Cherry, C., Dolan, W.B.: Data-driven response generation in social media. In: Proceedings of the Conference on Empirical Methods in Natural Language Processing, pp. 583–593. Association for Computational Linguistics (2011)
10. Schroff, F., Kalenichenko, D., Philbin, J.: FaceNet: a unified embedding for face recognition and clustering. CoRR abs/1503.03832 (2015). http://arxiv.org/abs/1503.03832
11. Serban, I.V., et al.: A deep reinforcement learning chatbot. arXiv preprint arXiv:1709.02349 (2017)
12. Tan, M., Xiang, B., Zhou, B.: LSTM-based deep learning models for non-factoid answer selection. CoRR abs/1511.04108 (2015). http://arxiv.org/abs/1511.04108
13. Wang, H., Lu, Z., Li, H., Chen, E.: A dataset for research on short-text conversations. In: Proceedings of the 2013 Conference on Empirical Methods in Natural Language Processing, pp. 935–945 (2013)
14. Wu, Y., Wu, W., Zhou, M., Li, Z.: Sequential match network: a new architecture for multi-turn response selection in retrieval-based chatbots. CoRR abs/1612.01627 (2016). http://arxiv.org/abs/1612.01627
15. Yan, Z., et al.: DocChat: an information retrieval approach for chatbot engines using unstructured documents. In: Proceedings of the 54th Annual Meeting of the Association for Computational Linguistics (Volume 1: Long Papers), vol. 1, pp. 516–525 (2016)
16. Zhou, X., et al.: Multi-view response selection for human-computer conversation. In: Proceedings of the 2016 Conference on Empirical Methods in Natural Language Processing, pp. 372–381 (2016)

A Model-Free Comorbidities-Based Events Prediction in ICU Unit

Tatiana Malygina[1,2]([⊠]) [iD] and Ivan Drokin[2]([⊠])

[1] The Laboratory of Bioinformatics, ITMO University,
Kronverkskiy pr. 49, St. Petersburg 197101, Russia
[2] Intellogic Limited Liability Company (Intellogic LLC), Office 1/334/63, Building 1,
42 Bolshoi blvd., Territory of Skolkovo Innovation Center, 121205 Moscow, Russia
{tanya.malygina,ivan.drokin}@botkin.ai
http://botkin.ai/

Abstract. In this work we focus on recently introduced "medical concept vectors" (MCV) extracted from electronic health records (EHR), explore in similar manner several methods useful for patient's medical history events prediction and provide our own novel state-of-the-art method to solve this problem. We use MCVs to analyze publicly-available EHR de-identified data, with strong focus on fair comparison of several different models applied to patient's death, heart failure and chronic liver diseases (cirrhosis and fibrosis) prediction tasks. We propose ontology-based regularization method that can be used to pre-train MCV embeddings. The approach we use to predict these diseases and conditions can be applied to solve other prediction tasks.

Keywords: Electronic health records · Ontology-based regularization
Neural networks · Health care · Learning (artificial intelligence)
Neural nets · Deep learning · Deep neural network
Electronic health record data · Data models · Diseases
Machine learning

1 Introduction

The numerous attempts to build recommendation systems for healthcare during recent decades were driven by its greatest problem - many patient deaths are caused by medical errors. The main goal of recommendation systems is to reduce harm to patient.

Typical system relies on a set of rules, which are manually-curated or extracted from medical data (or a combination of both). There are thousands of diseases, number of the system's rules is quite large, and our knowledge about it changes quite often. Hence, first manually-curated approach means a lot of work and sometimes leads to chaos. The second approach needs to apply various machine learning and data mining methods to medical data, and shifts the problem to ethics area: you need to ensure that you model learns common rules instead of learning to represent specific patients.

© Springer Nature Switzerland AG 2018
D. Ustalov et al. (Eds.): AINL 2018, CCIS 930, pp. 98–109, 2018.
https://doi.org/10.1007/978-3-030-01204-5_10

Moreover, patient's electronic health record (EHR) can contain heterogeneous data – from patient's diagnoses (given as a set of medical ontology codes) to CT scans (images) or ECG measurements (waveforms). Some information might be missing or incomplete.

The more patient's EHRs we get, the better we can predict their outcomes. It leads to the situation when millions of features (low-level data, i.e. 1-hot vectors for diagnoses, prescriptions or drugs) become not very effective for building precise models because of training speed and tendency to overfitting.

The clever solution for feature extraction in this case is called "medical concept vector" (MCV). The idea is quite similar to NLP embeddings: by projecting medical data to compressed space we obtain succinct representations capable to represent hidden relations between them.

1.1 Related Works

In recent decade the impressive results with NLP started when the development of technics for efficient representation of words in succinct vector space has begun with Word2vec model in 2 basic forms – continuous bag of words and skip-gram [11].

Since that, the idea of word embeddings was applied to various different tasks, including healthcare problems. To build biomedical embeddings the scientist needs to extract them from some data. For example, [12] use PubMed with MESH classification terms as a source of such data. The resulting embedding matrices represent co-occurrences between biomedical single-word terms. There are several human-accessed datasets to evaluate and compare them, including UMLS:Similarity package [10].

But what about patient's EHR data? It turns out that we are more interested in developing solutions for EHR-related tasks based on patients medical history. Its sparse representation is memory-consuming; model training for this model is also time-consuming. This problem is solved by building MCVs and using them to compress each patient's data. To represent relations between diagnoses and other medical events in patient history we prefer a co-occurrence matrix to single-word-based embeddings – these representations, called MCV, are more useful as a features for solving EHR-related predictions tasks like patient's outcomes, readmissions, length of stay in hospital, etc.

The extraction scheme for these embeddings is shown at Fig. 1. We consider a timeline of patient's admissions. Each admission is represented as a point with particular associated timestamp and various data, including diagnoses, procedures and prescriptions associated with it. We build a multi-hot representation of this timeline point, and then build sparse patient representation as a some form of aggregation for these vectors.

This basic scheme is modified for some cases. If we need to retain a sequence of diagnoses, we shuffle them within each admission for each batch. We do this to prevent overfitting and to emphasize that data associated with the same admission have the same timestamp and are unordered.

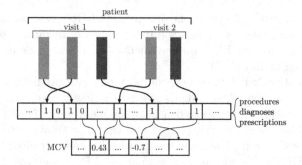

Fig. 1. Medical concept vectors extraction from patient's EHR. This typical scheme represent patient's medical history as a sequence of ordered visits, each of them has time mark. Every visit has associated diverse information - i.e., procedures, diagnoses and prescription codes. Each of these codes can be associated with 1-hot representation, together they turn patien't medical history to sparse multi-hot vector, which can be compressed to patient's embedding vector, which is called MCV.

Sometimes when data are limited we need to improve patient's MCV representation. Some approaches include but not limited to:

1. Use additional patient's data (procedural, prescription codes, etc.) [3,7].
2. Use common knowledge-based information (i.e., Clinical Classifications Software for ICD-9-CM or SNOMED-CT diagnoses codes ontology) [4].
3. Improve model by concatenation of patient's demography data to medical concept vector [5] or by utilizing temporal (or any other) admission information [7].

The simplest form of compression is achieved with the embedding matrix, which helps to project word representation given as a 1-hot encoded vector (with size equal to number of words in a vocabulary) to N-dimensional vector space.

We compare embeddings trained according to procedure defined by Choi et al. [4] which uses ICD9 hierarchy tree to produce embedding matrix.

They start with simple idea to incorporate knowledge DAG, composed from ICD-9 ontology tree, into the model and then use it to define positions between connected entities in embedding space (see Fig. 2).

The GRAM by [4] uses graph-based attention mechanism to learn MCV representations. In this model ith diagnosis vector l_i in MCV space becomes a convex combination of the basic embeddings e_j of itself and its ancestors [4]:

$$g_i = \sum_{j \in A(i)} \alpha_{ij} e_j, \quad \sum_{j \in A(i)} \alpha_{ij} = 1, \ \alpha_{ij} \geq 0 \text{ for } j \in A(i),$$

For each ICD-9 code it emphasizes its significant ancestors, and then defines particular code's MCV as weighted sum of ancestor MCVs and position of ICD-9 code in embedding space.

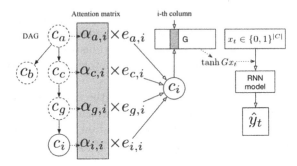

Fig. 2. GRAM model by [4]. In this model embeddings are represented in implicit form and are trained jointly with RNN, unlike our model, where we use pre-trained MCV embeddings.

2 Methods

2.1 Embedding with ICD9 Tree: Formalization

We follow the idea proposed by [4] to incorporate knowledge about ICD-9 code's ancestors in ICD-9 hierarchy tree, and expand it from different perspective. The authors of GRAM representation compute it as a part of RNN model and define an impact of ancestors at the training phase via attention mechanism.

Our approach is model-free, since we deduce comorbidities from data only and then use these embeddings as a pre-trained representations for diagnoses to improve predictive models.

Our regularizer contains several additive terms deduced from simple assumptions.

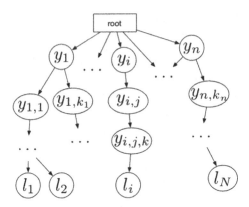

Fig. 3. ICD9 medical ontology scheme. This scheme is typical, however, it shows the notation used in our regularizer.

Suppose that we have a set of N diagnoses $V = \{l_1, l_2, \ldots, l_n\}$ mapped to medical ontology DAG with nodes denoted by y_i, where i is a sequence of indices describing path from ontology's root to node y_i (illustrated at Fig. 3). We call a 'leaf' an element of V, and 'node' is a medical ontology's category representing disease group.

It seems natural for leaves which belong to the same node to have similar MCV representations, and counterwise, leaves which belong to different nodes should have MCV representations which are distant in vector space. Thus, to get meaningful and comprehended MCV representations for diagnoses, we start from following assumptions:

- We follow the idea of skip-gram model: diagnoses which appear in patient's EHR timeline together, might be linked, thus distance between them in MCV space should be minimized.
- MCV representations for diagnoses with common ancestor in DAG should be closer than MCV representations for diagnoses with different ancestors in DAG.
- Distance between MCV representations for neighbouring nodes in MCV hierarchy should be closer to each other than distance between distant nodes. Distance between nodes and their common ancestor should be minimized.

For each diagnosis l_i we denote embedding MCV vector by $v(l_j)$, for each node y_i in medical ontology DAG we denote corresponding embedding MCV vector by $\tilde{v}(y_i)$. We use these to define a function

$$L_{K-DAG}(l_i) = \sum_{j \in N} ||v(l_i) - v(l_j)|| - \sum_{j \in T/N} ||v(l_i) - v(l_j)|| \qquad (1)$$

where N is a set of diagnoses with common parent node in DAG, and T/N is a set of diagnoses which ancestral node in DAG differs from the l_i's.

We minimize this function by minimizing a distance to l_i's neighbouring nodes in DAG and by maximizing a distance to everything distant.

For l_i we also minimize ontology-based term which is defined as follows. Lets suppose that l_i's parent node in DAG is $y = y_{i,j,k}$, then ontology-based regularization term is defined by

$$L_{NK-DAG}(y) = \sum_{t \in c(y_{i,j,k})} ||\tilde{v}(y_{i,j,k}) - v(l_t)|| \qquad (2)$$

$$+ \sum_{(\eta, \xi, \tau) \in n(y_{i,j,k})} ||\tilde{v}(y_{i,j,k}) - \tilde{v}(y_{\eta, \xi, \tau})|| \qquad (3)$$

$$- \sum_{(\eta, \xi, \tau) \in \overline{n}(y_{i,j,k})} ||\tilde{v}(y_{i,j,k}) - \tilde{v}(y_{\eta, \xi, \tau})|| \qquad (4)$$

$$+ \sum_{k \in \ldots} ||\tilde{v}(y_{i,j}) - \tilde{v}(y_{i,k})||$$

$$+ \sum_{\eta,\xi \in n(y_{i,j})} ||\tilde{v}(y_{i,j}) - \tilde{v}(y_{\eta,\xi})||$$

$$- \sum_{\eta,\xi \in \overline{n}(y_{i,j})} ||\tilde{v}(y_{i,j}) - \tilde{v}(y_{\eta,\xi})||$$

$$+ \sum_{j \in H} ||\tilde{v}(y_i) - \tilde{v}(y_{ij})|| - \sum_{\eta \in H/i} ||\tilde{v}(y_i) - \tilde{v}(y_\eta)||. \tag{5}$$

In the equation given above line Eq. (2) states that we minimize distance from specific node $y_{i,j,k}$ in medical ontology DAG to all its children (denoted in formula as $c(y_{i,j,k})$); we also minimize distance from this node to its neighbours $n(y_{i,j,k})$ in DAG (line Eq. (3)), and maximize distance to non-neighbour nodes at the same level of nodes hierarchy tree (line Eq. (4)). To write down this formula, we traverse the tree from $y_{i,j,k}$ to root node, composing the terms in similar manner. We stop after writing down terms for level 1 nodes (line Eq. (5)).

When we build this function for node located on different level of hierarchy, we compose it from all terms appearing during DAG traversal from y to root.

As a result, we define MCV embedding regularization by

$$\sum_{l_i} (L_{sy}(l_i) + L_{K-DAG}(l_i) + L_{NK-DAG}(y)), \tag{6}$$

where $L_{sy}(l_i)$ is a skip-gram's regularization for word l_i. We minimize this expression while training embeddings.

2.2 Dataset

To formulate health-related events prediction problem, we use data from MIMIC-III (Medical Information Mart for Intensive Care III) [8] dataset. These data are de-identified and freely available, they contain health information recorded for 40 thousand patients of Bet Israel medical center ICU unit from 2001 to 2012.

MIMIC-III de-identifies each patient, yet preserves information on patient's demography, related laboratory events and measurements, procedures, prescriptions, various patient-related events (including death) and waveforms, recorded during patient's ICU stays at medical center.

As you can see at Table 1, more than a half of de-identified patient records from MIMIC-III dataset has information only on 1 admission. This makes these records impossible to use for solving patient outcome prediction tasks, and reduces amount of useful data.

Typical ICU admission is longer than just one-time doctor visit. That leads to wide range of diagnoses counts associated to one ICU admission listed in database.

2.3 Benchmarking

Data Preprocessing. To enable comparison of several different models, we first prepared data using an approach similar to proposed at [7].

Table 1. Characteristics of MIMIC-III dataset

Demographics	
All patients	46518
Female	20398 (43%)
Admissions per patient	
0	–
≥ 1 and <2	39017
≥ 2 and <5	7221
≥ 5	280
Distinct admissions by types	
ELECTIVE	7696
EMERGENCY	42043
NEWBORN	7862
URGENT	1332
Admission 1-year windows (URGENT and EMERGENCY)	
1	3285
2	906
3	338
4	129
5	60
≥ 6 and <10	74
≥ 10	28

MIMIC-III contains a set of .csv relational-like tables, which are presented in normalized form. This means that for each patient's admission we can easily obtain a set of related information, including ICD9 diagnosis codes, prescriptions, medical procedures, and sort them according to admission's date for each patient.

There are 4 admission types available: URGENT, EMERGENCY, NEW-BORN and ELECTIVE. We excluded admissions with codes "NEWBORN" and "ELECTIVE", since there are few associated admission records, and these two types differ from "URGENT" and "EMERGENCY".

More than a half of patient records at MIMIC-III is associated with information about 1 or less admissions. We excluded these patients, because we cannot get labels used for training and validation for these patients and thus we cannot use them on diagnosis prediction tasks. We used these patients to solve different task - to build Word2vec-like MCV embedding matrices with hidden associations between various patient data.

We split each patient's admission history to independent samples using 1-year sliding window. We define a sample as sequence of at least 2 patient's admissions within 1 year.

For each sample, we used all related admissions except latest one for feature extraction, latest admission was used for label extraction. After that we splitted our dataset to train and test at 4:1 ratio.

MIMIC-III doesn't contain information related to events ordering within admission.

In proposed preprocessing method, different samples contain different number of admissions. We use zero-padding during extraction of features to unify lengths of different samples features.

Pipeline. For each particular prediction problem, we perform comparison of models with following steps:

1. **Construction of data split**, which is performed based on given stratification function, to ensure that both train and test datasets will contain positive and negative examples. After we choose indices for train and test datasets, we extract data for specific models. The representation of samples and corresponding labels might look different.
   We also do balancing by over-sampling for train dataset at this step to prevent classifiers of becoming constant. We do no balancing for test dataset.
2. We convert train and test datasets to the form compatible with each particular model, train each model and compare results. We do this in similar manner, i.e., we use the same number of epochs for neural network-based models, or we take into account the same number of recent diagnoses for all models, etc.

For each sample, which is given by sequence of patient's admissions within 1 year, we extract labels from last admission's data and use all preceding data for feature extraction.

Models. For each problem we compare the same set of classifiers. For a baseline, we compare MCV-based models with the following:

- **TF-IDF encoding**
  We process sparse multi-hot input vector of diagnoses with TF-IDF algorithm [9]. After that we use this matrix as an input to build logistic classifier.

For the following several models we use the same neural network architecture based on ResNet with attention layer, which utilize different MCV embedding matrices.

- **Word2Vec embeddings**
  For this model the embedding matrix was built from diagnoses, procedures and prescription co-occurrence patterns during single admission. We trained Word2vec with skip-gram mechanism to be able to transform given diagnosis, procedure or prescription to medical concept vector with size $N = 400$. We

picked output dimension size based on Edward Choi article [3] as the best value for hyperparameter. To obtain MCV representations, we use gensim [13].

After that we use this matrix to set weights in embedding layer at our basic model.

- **Word2Vec embedding+attention**

  We used the same model to train Word2Vec matrix as in the previous case, after that it used the same attention-based network architecture (with no data provided by second Input layer). To enable modifications, we implemented skip-gram model in keras [6] and added attention layer.

- **Embedding with ICD9 tree**

  To build embedding matrix-based model with ICD9 codes ontology we use modified skip-gram model with custom regularization defined in previous section.

  For each input patient, we build multi-hot representation of its medical history, after that we multiply it to MCV matrix. Thus we receive sum of all MCVs for all diagnoses appeared in patient's EHR.

  We use logistic classifier on patient's compressed representation to solve prediction tasks.

- **Embedding with ICD9 tree+attention**

  We use the procedure to construct embedding matrix as described above, after that we use construct neural network with attention mechanism.

- **Embedding with ICD9 tree+attention+tfidf**

  This approach differs from previous model with ICD-9 ontology-based embedding matrix and attention mechanism - before making prediction, we concatenate to patient's representation the output of **TF-IDF encoding** model, provided to second Input layer.

- **Choi embedding+attention**

  For comparison, we trained MCV embeddings matrix with the method proposed by [2,4]. We run RNN-model till convergence. After that we extracted from it diagnoses representation described as G matrix at [4].

  We used this matrix G to set weights in embedding layer at our model. We didn't provide data to additional Input layer and trained it as all previous models – on the same training dataset.

Time-Based Model

We also reproduce method for medical concept vector computation from temporal perspective as described by [7]. To build embedding matrix on diagnoses, prescriptions and procedures data we use gensim package by [13].

3   Results

We trained models to solve 3 different tasks: death prediction, heart failure and chronic liver diseases. For each problem, we trained all models and compared results.

Table 2. Predictions

	Accuracy	Precision	Recall	AUROC
Mortality prediction				
TF-IDF encoding	0.718	0.500	0.519	0.732
Choi embedding+attention	0.795	0.648	**0.597**	0.769
ICD9 tree	0.692	0.462	0.558	0.677
ICD9 tree+attention	0.780	0.635	0.519	0.747
ICD9 tree+attention+tfidf	**0.799**	**0.672**	0.558	**0.799**
Time based model	0.736	0.531	0.558	0.764
Word2Vec	0.670	0.429	0.506	0.658
Word2Vec+attention	0.777	0.621	0.532	0.779
Heart failure predictions				
TF-IDF encoding	0.846	0.874	0.810	0.918
Choi embedding+attention	0.828	0.875	0.766	0.909
ICD9 tree	**0.857**	0.866	**0.847**	0.908
ICD9 tree+attention	0.821	0.855	0.774	0.920
ICD9 tree+attention+tfidf	0.842	**0.892**	0.781	**0.924**
Time based model	0.751	0.732	0.796	0.854
Word2Vec	0.766	0.783	0.737	0.832
Word2Vec+attention	0.806	0.850	0.745	0.906
Liver diseases				
TF-IDF encoding	0.996	1.000	0.875	0.998
Choi embedding+attention	0.996	0.889	1.000	1.000
ICD9 tree	0.982	0.615	1.000	0.998
ICD9 tree+attention	**1.000**	**1.000**	**1.000**	**1.000**
ICD9 tree+attention+tfidf	0.996	0.889	1.000	1.000
Time based model	0.989	0.778	0.875	0.992
Word2Vec	0.971	0.500	0.875	0.995
Word2Vec+attention	0.989	0.727	1.000	1.000

For each particular prediction task we split dataset using the procedure described at Benchmarking section, to train every model on the same data and to compare results on the same test dataset.

We define labels for each prediction task based on data found in latest visit within 1-year of patient's admission.

For death prediction task we say that label is 1 if patient dies during latest visit and 0 if otherwise. Frequency of the majority class was around 0.75 (this can be considered as a baseline on test dataset).

For heart failure we define label to be equal 1 if patient's set of diagnoses during latest visit includes any diagnoses corresponding to "heart failure"[1] Frequency of the majority class was around 0.50 (this can be considered as a baseline on test dataset for this prediction task).

For liver diseases we define label to be equal 1 if set of diagnoses during latest visit contains one of the following (given as corresponding ICD-9 codes from MIMIC-III subset of diagnoses): 57.12, 57.15, 57.16, and to be equal 0 if otherwise. Frequency of the majority class was around 0.95 (this can be considered as a baseline on test dataset).

The result of comparison is shown at Table 2. We balanced data by oversampling during training, but used no balancing while testing results.

4   Discussion

The main problem we had to deal with was lack of de-identified freely-available datasets. One might argue that synthetic datasets might be used instead. But they don't take into consideration patient's demography, and are likely to produce unreliable embeddings being used as a training dataset (as a particular example, in one of them authors of this article found record for pregnant 70-year old men with HIV and cancer).

Impressive results shown at [1,3] can hardly be reproduced, since they were achieved on private datasets 5–10 times bigger than publicly available de-identified MIMIC-III.

Despite that we managed to build succinct representations and show that our model provides better results on several prediction tasks.

References

1. Avati, A., Jung, K., Harman, S., Downing, L., Ng, A., Shah, N.H.: Improving palliative care with deep learning. In: 2017 IEEE International Conference on Bioinformatics and Biomedicine (BIBM), pp. 311–316, November 2017. https://doi.org/10.1109/BIBM.2017.8217669
2. Choi, E., Bahadori, M.T., Schuetz, A., Stewart, W.F., Sun, J.: Doctor AI: predicting clinical events via recurrent neural networks. In: Doshi-Velez, F., Fackler, J., Kale, D., Wallace, B., Wiens, J. (eds.) Proceedings of the 1st Machine Learning for Healthcare Conference, Proceedings of Machine Learning Research, vol. 56, pp. 301–318. PMLR, Children's Hospital LA, Los Angeles, CA, USA, 18–19 August 2016. http://proceedings.mlr.press/v56/Choi16.html
3. Choi, E., et al.: Multi-layer representation learning for medical concepts. In: KDD, pp. 1495–1504 (2016)
4. Choi, E., Bahadori, M.T., Song, L., Stewart, W.F., Sun, J.: GRAM: graph-based attention model for healthcare representation learning. In: Knowledge Discovery and Data Mining (KDD) (2017)
5. Choi, Y., Chiu, C.Y.I., Sontag, D.: Learning low-dimensional representations of medical concepts. AMIA Jt. Summits Transl. Sci. Proc. **2016**, 41–50 (2016)

---

[1] Corresponding ICD-9 codes are: 398.91, 402.01–402.91, 404.01–404.93, 428.0–428.43.

6. Chollet, F., et al.: Keras (2015). https://github.com/keras-team/keras
7. Farhan, W., Wang, Z., Huang, Y., Wang, S., Wang, F., Jiang, X.: A predictive model for medical events based on contextual embedding of temporal sequences. JMIR Med. Inf. **4**, e39 (2016). https://www.ncbi.nlm.nih.gov/pmc/articles/PMC5148810/
8. Johnson, A., et al.: MIMIC-III, a freely accessible critical care database. Sci. Data **3**, 160035 (2016). https://doi.org/10.1038/sdata.2016.35. http://www.nature.com/articles/sdata201635
9. Jones, K.S.: A statistical interpretation of term specificity and its application in retrieval. J. Doc. **60**(5), 493–502 (2004)
10. McInnes, B.T., Pedersen, T., Pakhomov, S.V.: UMLS-interface and UMLS-similarity: open source software for measuring paths and semantic similarity. In: AMIA Annual Symposium Proceedings, pp. 431–435 (2009)
11. Mikolov, T., Chen, K., Corrado, G., Dean, J.: Efficient estimation of word representations in vector space. In: Proceedings of Workshop at ICLR, vol. 2013, January 2013
12. Pyysalo, S., Ginter, F., Moen, H., Salakoski, T., Ananiadou, S.: Distributional semantics resources for biomedical text processing. In: Proceedings of LBM, pp. 39–44 (2013)
13. Řehůřek, R., Sojka, P.: Software framework for topic modelling with large corpora. In: Proceedings of the LREC 2010 Workshop on New Challenges for NLP Frameworks, pp. 45–50. ELRA, Valletta, Malta, May 2010. http://is.muni.cz/publication/884893/en

Explicit Semantic Analysis as a Means for Topic Labelling

Anna Kriukova[1(✉)], Aliia Erofeeva[2], Olga Mitrofanova[1], and Kirill Sukharev[3]

[1] St. Petersburg State University, St. Petersburg, Russia
krukova.ann@gmail.com, oa-mitrofanova@yandex.ru
[2] University of Trento, Trento, Italy
amirzagitova@gmail.com
[3] St. Petersburg Electrotechnical University, St. Petersburg, Russia
sukharevkirill@gmail.com

Abstract. This paper deals with a method for topic labelling that makes use of Explicit Semantic Analysis (ESA). Top words of a topic are given to ESA as an input, and the algorithm yields titles of Wikipedia articles that are considered most relevant to the input. An alternative approach that serves as a strong baseline employs titles of first outputs in a search engine, given topic words as a query. In both methods, obtained titles are then automatically analysed and phrases characterizing the topic are constructed from them with the use of a graph algorithm and are assigned with weights. Within the proposed method based on ESA, post-processing is then performed to sort candidate labels according to empirically formulated rules. Experiments were conducted on a corpus of Russian encyclopaedic texts on linguistics. The results justify applying ESA for this task, and we state that though it works a little inferior to the method based on a search engine in terms of labels' quality, it can be used as a reasonable alternative because it exhibits two advantages that the baseline method lacks.

Keywords: Topic labels · Topic modelling
Explicit Semantic Analysis · Russian

1 Introduction

One of the most claimed approaches in contemporary computational semantics is topic modelling which describes a corpus in terms of latent topics and reveals the distribution of documents over topics. Being a variety of fuzzy clustering, such representation characterises text semantics and effectively depicts a structure of large document collection. Researchers have developed various types of topic models, preferences being given to Probabilistic Latent Semantic Analysis (pLSA) [5] and Latent Dirichlet Allocation (LDA) [3]. In this study we focus our attention to LDA which is a generative model consisting of two stages: (1) distribution θ_d of documents d over topics t in a collection D is defined; (2) topic ϕ_t

D. Ustalov et al. (Eds.): AINL 2018, CCIS 930, pp. 110–116, 2018.
https://doi.org/10.1007/978-3-030-01204-5_11

for a word w in a document d is chosen in accordance with distribution θ_d. It is supposed that θ_d and ϕ_t conform with distributions $Dir(\alpha)$ and $Dir(\beta)$ where α and β are considered as hyperparameters of Dirichlet allocation. In practice, the number of topics and their size are defined by users in course of experiments.

Generated topics are standardly presented to end-users as a list of the top n terms from the multinomial distribution of words ranked by the probability $Pr(w|\phi_t)$. However, this often hinders the understanding of a topic by human readers, especially when the selected number of topics is large. Aimed to reduce the cognitive load of interpreting such topics, the task of automatic topic labelling emerged [10], i.e. the task of finding a concise and salient label that describes the content of a given topic. The problem has been studied extensively for English, and there have been proposed numerous methods of topic labelling, varying in label modality (words [9] and phrases [2,8,10], or images [1], or both [16]), label generation (relying only on the content of the modelled corpus [6,9,10] or involving external resources [2,8,16]), and algorithms employed (broadly, supervised [1,8,16] or unsupervised [2,6,9,10]).

This line of research is still actively developing in Russian NLP: a label that generalises words of a topic would make its interpretation substantially easier. Our project continues experiments in this field. Therefore, the main task of our paper is to perform comparative analysis of two algorithms adjusted for topic labelling: Explicit Semantic Analysis which relies upon external knowledge from Wikipedia, and Graph-based topic labelling which implies label extraction from a search engine output.

2   Graph-Based Topic Labelling

As a strong baseline topic labelling algorithm, we take the unsupervised graph-based method first introduced for English [2] and proved to be applicable to Russian [12,13]. In the following, we briefly describe the procedure.

At the initial stage of candidate generation, the first k topic words are used as a single query to a search engine. The titles of the top n search results are stripped from stopwords and concatenated into a continuous synthetic text which is then lemmatised and fed into the TextRank [11] ranking algorithm. The text is transformed into an oriented graph $G = \{V, E\}$, where V is a set of nodes representing lemmata, E is a set of weighted edges defined by some similarity metric, e.g. the co-occurrence frequency within the input text. Next, the TextRank value is recursively computed for each node based on the in- and out-degrees [11]. Nodes (words) having higher scores are assumed to be more salient, while edges with larger weights indicate a stronger semantic association between the corresponding word pairs.

In order to move from single tokens to higher level n-grams, the algorithm has been tailored for Russian by applying a set of manually crafted morphological patterns to extract grammatically valid key phrases [12]. At the stage of candidate selection, having each lemma assigned a TextRank score allows ranking the phrases according to the sums of weights of the constituent words.

3   Explicit Semantic Analysis

The approach we concentrate on in this paper makes use of the algorithm for constructing topic labels described in Sect. 2, except that we employ Explicit Semantic Analysis (ESA) [4] rather than a search engine as a way for using external knowledge sources. ESA is a way of representing words and texts in a vector space. ESA makes use of a large collection of documents as a knowledge source: the authors of the initial paper carried out experiments on Wikipedia[1] and described how ESA can be used for both mono- and cross-lingual tasks. Wikipedia is an open and constantly growing source of Russian texts (the Russian Wikipedia contains now almost 1.5 mln articles). Wikipedia articles in ESA are treated as *concepts*, because each article is supposed to describe in detail a single topic. The algorithm deploys the "bag-of-words" approach for representation of concepts, which is often used for NLP problems. Though being a simplification of real-world intertextual relations, it is justified here by the fact that we can usually describe any concept by means of separate words associated with it. Each concept therefore is described by a vector that contains words co-occurring in the corresponding article. Words are assigned with TF-IDF [15] weights that reflect association strength. ESA thus represents text meaning in terms of a weighted concept vector, sorted according to the relevance of concepts to the text. At this point an inverted index is created, that bounds a word with concepts where it occurs. If a concept's weight for a given word is too small, the concept is deleted from the interpretation vector, which allows us to eliminate insignificant links between words and concepts. The intuition behind such representations of text semantics is that in this way we get the most important concepts related to a text and can represent its meaning with their help.

Summing up, ESA represents meaning of a text in a high-dimensional space of concepts derived from Wikipedia. A vector for topic words contains TF-IDF values, i.e. figures; however, as we know which number refers to which article, we can now get a list of articles' titles sorted in the descending order by their weights. The titles are then processed in the same way as search engine results in the previously described algorithm.

4   Experiments

Experiments were performed on topics obtained from a corpus of Russian encyclopaedic texts on linguistics [12]. Size of the corpus, containing more than 1.3 million tokens before pre-processing, reduced to 934,855 tokens after lemmatisation and removing of stop words, digits, and punctuation. We extracted 20 topics using the LDA model from *scikit-learn* [14], with default settings.

Top 10 words from each topic were used as an input for ESA and Yandex[2] search engine, and given the titles of 30 most relevant Wikipedia articles from ESA and search results from Yandex, we then applied the procedure described

---

[1] https://www.wikipedia.org.
[2] https://yandex.ru.

Table 1. Example of ESA output, titles discarded in post-processing are shown in red.

Topic: время лексема реконструкция том часы число средний антоним фигура использоваться ("time lexeme reconstruction volume hours number neutral antonym figure being_used")	Лексема, Реконструкция, 48 часов, Часы (значения), GiST, Токен, Мариус (значения), T-15 ("Lexeme, Reconstruction, 48 hours, Hours (disambiguation), GiST, Token, Marius (disambiguation), T-15")

in Sect. 2: we created a graph and weighted the candidate labels using the TextRank algorithm. The methods based on the search engine and on ESA will be referred to below as Labels-Yandex and Labels-ESA, respectively, for the sake of convenience.

The initial results provided by Labels-ESA turned out, however, to be rather noisy. First of all, after analyzing intermediate ESA outputs (when it is presented with topic words), we decided to exclude the following article titles (ref. Table 1):

1. dedicated to people, as names would not likely make a meaningful label;
2. containing numbers, for numbers are not supposed to serve as a topic label;
3. containing words not in Russian, since they are later deleted as stop-words;
4. whose length is less than 3 symbols (e.g. articles about alphabet letters);
5. containing the mark "(значения)" ("disambiguation"), as such titles refer to pages with links to other articles.

We also revealed some characteristic features of ESA, probably because of which good labels did not end up at top positions in labels lists and the algorithm needed some enhancement. Firstly, top Wikipedia articles that are delivered by ESA sometimes seem to characterise only few topic words out of ten. For example, if a topic contains the word "диалект" ("dialect"), the first articles describe only different kinds of dialects. That is connected with the manner how a text vector is formed within the ESA approach. ESA combines vectors for each word in a text, i.e. ten vectors for topic words in our case. Thus, if a particular topic word has high TF-IDF values for its articles, they tend to outweigh other words articles in the text vector resulting in its being almost the same as for this only word.

Secondly, ESA finds hyponyms of words more likely than hyperonyms. For example, for a word "гласный" ("vowel") ESA would find articles like "гласныйпереднего ряда верхнего подъёма" ("high front vowel"), rather than articles like "звуки" ("sounds") or "фонетика" ("phonetics"). It takes place as in specific articles, words we are trying to characterise are normally mentioned more often than in general articles, whereas for making topic labels, hyperonyms are more likely required. Thirdly, it is homonymy and polysemy. ESA does not take it into account when searching for most relevant articles because words in the Wikipedia dump are not provided with such information. Thus, some articles in the output can actually be connected with other domains.

Taking all these into account, we decided to manually write rules that would rearrange lists with 20 first labels by Labels-ESA so that the relevant ones would be drawn up. The following post-processing rules were generated empirically:

Table 2. Top-3 labels assigned to some of the topics by Labels-ESA and Labels-Yandex.

Topic	Labels-ESA	Labels-Yandex
лингвистика язык наука лингвистический теория метод исследование анализ идея год ("linguistics language science linguistic theory research analysis idea year")	история лингвистики, морфологический анализ, словари лингвистических терминов ("history of linguistics, morphological analysis, dictionaries of linguistic terms")	методы лингвистического анализа, методология лингвистического анализа, лингвистический анализ ("methods of linguistic analysis, methodology of linguistic analysis, linguistic analysis")
форма глагол язык вид значение время действие наклонение наречие иметь ("form verb language aspect meaning tense act modality adverb have")	спряжение глаголов, наклонение, лингвистика ("verb conjugation, modality, linguistics")	категория наклонения глагола, категория наклонения, наречие слова категории ("category of verb modality, category of modality, adverb category words")

1. If a two-word label is a part of a longer label, the latter is excluded and the former is moved to the first place (e.g. "statistical machine translation" → "machine translation").
2. (a) If more than five labels contain the same noun, all of them are deleted and the noun in plural form is placed at the first position (e.g. the word "dialects" replaces different kinds of dialects).
   (b) If labels contain adjectives from the corresponding topic, we add the most frequent adjective to the noun from the previous step and also move the resulting label to the first place in the labels list.
3. If a label contains more than three words, it is moved back by two positions.
4. If a label contains more than one word and an adjective from the corresponding topic, it is placed at the first position.

The rules proved to considerably improve output of Labels-ESA. Some topics and their top three labels can be seen in Table 2.

5   Evaluation and Analysis

First of all, in the task of topic labelling there is a problem with evaluation, because no gold standard is usually available. Therefore, to evaluate the results we asked six experts to rate obtained labels manually. The experts, students at the department of mathematical linguistics at the St Petersburg State University, were to look through the first ten words for each topic and choose which of the top label from Labels-ESA and one the label from Labels-Yandex matches the corresponding topic better. It was also allowed to mark both or none of the labels. A part of the assessment can be seen in Table 3. We computed the mean value for each method (considering each plus as 1 and each minus as 0), which turned out to be 0.47 for Labels-ESA and 0.54 for Labels-Yandex.

Table 3. Evaluation examples of Labels-ESA and Labels-Yandex, in respective order.

Topic: перевод словарь текст система компьютерный машинный язык приклад-ной русский ("translation dictionary text system computational machine language applied Russian")	
машинный перевод ("machine translation")	+ − + + + +
система компьютерного перевода ("computer translation system")	+ + + − − +
Topic: язык литературный диалект русский говор современный норма немецкий ("language literary dialect Russian accent modern norm German")	
диалекты ("dialects")	+ + + + + +
диалекты немецкого языка ("German dialects")	− − − − − +
Topic: лингвистика язык наука теория метод исследование анализ идея год ("linguistics language science theory method research analysis idea year")	
история лингвистики ("history of linguistics")	− − + − − −
методы лингвистического анализа ("methods of linguistic analysis")	+ + − + + +

Although Labels-ESA is assessed a little worse, there are several reasons why using ESA instead of a search engine may be beneficial. First of all, when trying to automatically obtain titles from a search engine, one can run into a problem that there is often a limit set up on the number of queries per minute from one IP address when addressing a search engine automatically. Consequently, it takes much more time to find titles for a topic and assign it with a label. ESA has no such limitation and processes a ten-word topic at about 1.2 s. Secondly, search systems usually use complex algorithms for ranking pages, and, what is more, can individualize results for a certain user, which is why experiments of a method relying on a search engine may be hard to reproduce. In case of the ESA approach, we make use of a certain Wikipedia dump, so that the results remain consistent.

Both these characteristics let us regard ESA as a reasonable alternative to the baseline method.

6 Conclusions

In this work we propose to use Explicit Semantic Analysis as a means for dealing with automatic topic labelling. ESA has only recently been adopted for the Russian language and has yet been used for measuring degree of texts' semantic relatedness [7], while we describe its advantages and drawbacks with regards to topic labelling.

We compared our method, based on ESA, with an alternative algorithm that uses titles of first outputs in a search engine, given topic words as a query [12]. The work of both of them was evaluated on topic models extracted from a corpus of Russian encyclopaedic texts on linguistics. The evaluation procedure showed that our method works almost as well as the alternative algorithm, whereas it has a number of significant advantages. Future work will address assessing results on other corpora and domains to prove that post-processing of Labels-ESA output provides equally good results on different kinds of data.

References

1. Aletras, N., Mittal, A.: Labeling topics with images using a neural network. In: Jose, J.M., et al. (eds.) ECIR 2017. LNCS, vol. 10193, pp. 500–505. Springer, Cham (2017). https://doi.org/10.1007/978-3-319-56608-5_40
2. Aletras, N., Stevenson, M., Court, R.: Labelling topics using unsupervised graph-based methods. In: Proceedings of the 52nd Annual Meeting of ACL, pp. 631–636. ACL (2014). https://doi.org/10.3115/v1/P14-2103
3. Blei, D., Ng, A., Jordan, M.L.: Latent dirichlet allocation. J. Mach. Learn. Res. **3**, 993–1022 (2003). https://doi.org/10.1162/jmlr.2003.3.4-5.993
4. Gabrilovich, E., Markovitch, S.: Computing semantic relatedness using Wikipedia-based explicit semantic analysis. In: IJCAI International Joint Conference on Artificial Intelligence, pp. 1606–1611 (2007). https://dl.acm.org/citation.cfm?id=1625535
5. Hofmann, T.: Probabilistic latent semantic indexing. In: Proceedings of the 22nd Annual International ACM SIGIR Conference on Research and Development in Information Retrieval, pp. 50–57 (1999). https://doi.org/10.1145/312624.312649
6. Kou, W., Li, F., Baldwin, T.: Automatic labelling of topic models using word vectors and letter trigram vectors. In: Zuccon, G., Geva, S., Joho, H., Scholer, F., Sun, A., Zhang, P. (eds.) AIRS 2015. LNCS, vol. 9460, pp. 253–264. Springer, Cham (2015). https://doi.org/10.1007/978-3-319-28940-3_20
7. Kriukova, A., Mitrofanova, O., Sukharev, K., Roschina, N.: Using explicit semantic analysis and Word2Vec in measuring semantic relatedness of Russian paraphrases. In: 2018 Digital Transformations and Modern Society (2018)
8. Lau, J.H., Grieser, K., Newman, D., Baldwin, T.: Automatic labelling of topic models. In: Proceedings of the 49th Annual Meeting of the ACL, pp. 1536–1545. ACL, Stroudsburg (2011)
9. Lau, J.H., Newman, D., Karimi, S., Baldwin, T.: Best topic word selection for topic labelling. In: Proceedings of the 23rd International Conference on Computational Linguistics (COLING 2010), No. August, pp. 605–613 ACL, Stroudsburg (2010)
10. Mei, Q., Shen, X., Zhai, C.: Automatic labeling of multinomial topic models. In: Proceedings of the 13th ACM SIGKDD Knowledge Discovery and Data Mining, KDD 2007, p. 490. ACM Press (2007). https://doi.org/10.1145/1281192.1281246
11. Mihalcea, R., Tarau, P.: TextRank: bringing order into texts. In: Proceedings of EMNLP, vol. 85, pp. 404–411 (2004). https://doi.org/10.3115/1219044.1219064
12. Mirzagitova, A., Mitrofanova, O.: Automatic assignment of labels in topic modelling for Russian corpora. In: Botinis, A. (ed.) Proceedings of the 7th Tutorial and Research Workshop on Experimental Linguistics, pp. 107–110. ISCA, Saint Petersburg (2016). https://www.researchgate.net/publication/320444549
13. Panicheva, P., Mirzagitova, A., Ledovaya, Y.: Semantic feature aggregation for gender identification in Russian Facebook. In: Filchenkov, A., Pivovarova, L., Žižka, J. (eds.) AINL 2017. CCIS, vol. 789, pp. 3–15. Springer, Cham (2018). https://doi.org/10.1007/978-3-319-71746-3_1
14. Pedregosa, F., et al.: Scikit-learn: machine learning in python. J. Mach. Learn. Res. **12**, 2825–2830 (2011)
15. Salton, G., McGill, M.J.: Introduction to Modern Information Retrieval. McGraw-Hill, New York (1983)
16. Sorodoc, I., Lau, J.H., Aletras, N., Baldwin, T.: Multimodal topic labelling. In: Proceedings of the 15th Conference of EACL, vol. 2, pp. 701–706 (2017). https://doi.org/10.18653/v1/E17-2111

Four Keys to Topic Interpretability in Topic Modeling

Andrey Mavrin[1], Andrey Filchenkov[1(✉)], and Sergei Koltcov[2]

[1] ITMO University, St. Petersburg, Russia
andreyshambala@gmail.com, afilchenkov@corp.ifmo.ru
[2] National Research University Higher School of Economics, St. Petersburg, Russia
skoltsov@hse.ru

Abstract. Interpretability of topics built by topic modeling is an important issue for researchers applying this technique. We suggest a new interpretability score, which we select from an interpretability score parametric space defined by four components: a splitting method, a probability estimation method, a confirmation measure and an aggregation function. We designed a regularizer for topic modeling representing this score. The resulting topic modeling method shows significant superiority to all analogs in reflecting human assessments of topic interpretability.

Keywords: Topic modeling
Additive regularization for topic modeling · Topic interpretability
Human assessment

1 Introduction

Topic modeling is a domain of machine learning that has been actively developing since the late 1990s. Its main goal is to determine given a set of text documents, to which topics each document relates, as well as what terms each topic consists of. Topic modeling allows effectively solving of such tasks as clustering and classification of text documents [19], topical search of documents and related objects [17], building of topical profiles of users of various Internet resources [9], analysis of news flows [11] and many others.

In many cases, in the above-mentioned areas of topic modeling application requires a person to interact directly with the topic model. In these cases, the concept of "topic" has to correspond to the human notion of it. In particular, words that form a specific topic must be semantically related. The task of assessing the topic interpretability in topic models has been actively studied since the end of 2010, when the methods of expert assessment of interpretability [4] were first proposed, and later interpretability scores were suggested [1,12,14].

The goal of this work is improving interpretability of topics. To do so, we use additive regularization for topic modeling (ARTM) approach by proposing a regularizer that supports topic interpretability. For this purpose, we explore

D. Ustalov et al. (Eds.): AINL 2018, CCIS 930, pp. 117–129, 2018.
https://doi.org/10.1007/978-3-030-01204-5_12

interpretability scores in an interpretability score parametric space and find the one, which is the best to reflect human assessment of topic interpretability.

The rest of the paper is structured as follows. In Sect. 2, we briefly describe ARTM approach and several regularizers, with which we will compare our work. In Sect. 3, we describe the parametric space of interpretability score, as well as present a regularizer corresponding to such space. In Sect. 4 we briefly describe details of the method implementation and experimental evaluation. Results and their discussion is presented in Sect. 5. Section 6 concludes.

2   Related Work

2.1   Topic Modeling and Additive Regularization

The probabilistic topic model (TM) of a document collection is a set of topics, each of which is a probability distribution on the set of words encountered in the collection, and a set of probability distributions on a set of topics for each document [20]. Since the notation in topic modeling domain has not been changed during recent years, and the size of paper is limited, we will skip the notation assuming that a reader is familiar with it. We will follow [5,21,22].

Many approaches for topic modeling were suggested: Latent Semantic Analysis (LSA) [16], Probabilistic Latent Semantic Analysis (PLSA) [8], Latent Dirichlet Allocation (LDA) [2]. They were generalized under an approach suggested in 2014 by Konstantin Vorontsov [21] called additive regularization of topic models (ARTM). The main idea of this approach is to maximize model likelihood jointly with additional criteria called regularizers that represent additional constraints.

2.2   Topic Interpretability

Interpretability of topics obtained as the result of topic modeling began to be actively considered in 2009, when a method for assessing the interpretability of the topic by a person called word intrusion was proposed [4]. Intuitively, the assessment of topic interpretability is whether a person can understand how the words representing a topic are related to each other and what is a general concept to which they relate. The word intrusion method evaluating of the topic interpretability by a respondent is as follows. Each topic is presented in the form of six words, five of which are the most probable words in the topic, and the sixth word is chosen randomly from words in this topic having a low probability. The task of the respondent is to correctly determine the intruder. The interpretability of the topic is estimated by the number of respondents who found the intruder.

Due to the assessing of the topic interpretability is a very expensive and time-consuming procedure, it would be desirable to be able to evaluate interpretability without human participation. Researchers have suggested several scores for estimating the topic interpretability discussed below.

Pointwise Mutual Information. Idea of this score (more well-known as UCI) [13] is to assess the topic interpretability by associating all the pairs of words in a topic. Such association is estimated on some large external corpus. It is assumed that the topic is represented by the ten most likely words in this topic. The formula of the topic interpretability is as follows:

$$PMIScore(w) = \text{median}\{PMI(w_i, w_j), i, j \in \{1..10\}\},$$

$$PMI(w_i, w_j) = \log \frac{p(w_i, w_j)}{p(w_i)p(w_j)},$$

where $p(w)$ is word probability estimated on an external corpus, $p(w_i, w_j)$ is a joint probability of a pair of words estimated with a sliding window of size 10 scanning the external corpus.

UMass. This score [12] is quite similar to the UCI, however in this case the function estimating the association between a pair of words is not symmetric. In addition, it does not use external corpus, evaluating the coherence of words in the collection of documents on which TM was built. It is also assumed that a topic is described with M of its most probable words. It is defined as follows:

$$C(t, V^{(t)}) = \sum_{m=2}^{M} \sum_{l=1}^{m-1} \log \frac{D(v_m^{(t)}, v_l^{(t)}) + 1}{D(v_l^{(t)})},$$

where $D(v)$ is frequency of word v among the documents, $D(v, v')$ is the joint frequency of pair (v, v') among documents, $V^{(t)} = (v_1^{(t)}, \ldots, v_M^{(t)})$ is a list of M most probable words in topic t. The unit in the numerator under the logarithm prevents the value under the logarithm from being converted to zero.

Context Vectors. The main idea of this score [1] is usage of vector representation of words in the subject. It is also assumed that a topic is represented with n most probable words. The proposed score is defined as:

$$Coherence_{Sim}(T) = \frac{\sum_{1 \leq i \leq n-1, i+1 \leq j \leq n} Sim(\boldsymbol{w}_i, \boldsymbol{w}_j)}{\binom{n}{2}}$$

$$Sim(\boldsymbol{w}_i, \boldsymbol{w}_j) = \frac{\boldsymbol{w}_i \cdot \boldsymbol{w}_j}{||\boldsymbol{w}_i|| \cdot ||\boldsymbol{w}_j||},$$

where $\boldsymbol{w}_i$ is a vector representation of word $w_i \in T$. The vector representation is learned on an external corpus with so-called word context, which is defined as 10 words closest to each of the word occurrences into the outer body (5 on each side). Thus, every occurrence of word w in the chosen external corpus results in 10 new components in the vector representation of w. Value of w component associated with word f is evaluated as $PMI(w, f)^\gamma$, where γ is the parameter that makes the components of the vector with a high value to be more meaningful.

However, it is easy to see that with this approach the dimensionality of the vectors turns out to be too large, therefore it is suggested to limit the dimension by choosing only β_{w_i} of the most connected (maximal) components, where β_{w_i} is computed using the following formula [10]:

$$\beta_{w_i} = (\log(c(w_i)))^2 \cdot \frac{\log_2(m)}{\delta},$$

where δ is a regularization coefficient and m is the external corpus size.

2.3   Interpretability in ARTM

A regularizer for ARTM is known, which directly maximizes the coherence between words in a topic [22]. It uses a previously computed matrix of connectivity between the words C, where C_{uv} is the joint estimate of pair of words $(u, v) \in Q \subset W^2$. This regularizer, which minimizes the sum of divergences between each distribution of ϕ_{vt} and its estimate for all words that occur with v, looks like this:

$$R(\Phi) = \tau \sum_{t \in T} \sum_{(u,v) \in Q} C_{uv} n_{ut} \ln \phi_{vt} \to \max.$$

However, application of this regularizer meets some difficulties. Given a sufficiently large volume of the collection, on which topic model is built, it is not possible to evaluate the joint occurrence for each pair of words in the collection due to the very large size of the set of all pairs of words. A choice of some subset of pairs of words must have some logical justification, which also causes difficulties. This is why we did not include this approach in comparison.

Next modification is *word embedding coherence* (WEC) [15], which is:

$$\mathrm{Coh}_{we}(t) = \frac{1}{n(n-1)} \sum_{\tilde{w}_i{}^{(t)} \neq \tilde{w}_j{}^{(t)}} d(v(\tilde{w}_i{}^{(t)}), v(\tilde{w}_j{}^{(t)})),$$

where $v : \mathcal{W} \to \mathbb{R}^d$ is a mapping from tokens to d-dimensional vectors and $d : \mathbb{R}^d \times \mathbb{R}^d \to \mathbb{R}$ is a distance function.

3   Interpretability Scores and an Additive Regularizer

First, we describe a parametric space of interpretability scores, in which we will further search for the best metric. Second, we present a new regularizer for TMs, which maximizes the interpretability of the main words in topics.

3.1   Parametric Space of Interpretability Scores

We assume that a score estimating the quality of topic interpretability can be represented in the form of four relatively independent components [18] $(\mathscr{S}, \mathscr{P}, \mathscr{C}, \mathscr{A})$, which will be described in detail later. The input of a interpretability score are a topic and n of the most probable words $W =$

$\{w_1, w_2, \ldots, w_n\}$. The first component $\mathscr{S}$ of the score is a method of splitting the most probable words into pairs (W', W^*), where $W', W^* \subset W$. The second component $\mathscr{P}$ is a method of estimating word **probability**, which is a function $P : W' \rightarrow [0, 1]$. It is computed using an external collection of documents, which differs from the one, on which the topic model is built. Intuitive requirement for this collection is a presence of large amount of non-specific information. An example of such a collection is a set of Wikipedia articles. The third component $\mathscr{C}$ of the score is function $C : (W', W^*) \rightarrow \mathbb{R}$, which is the so-called **confirmation** measure. It shows how much the subset W' supports the subset W^*. The fourth component $\mathscr{A}$ is an **aggregating** function that converts a set of real numbers into single real number.

Thus, the whole process of computing interpretability score of a topic can be described as follows. First, topic W is split into a set of pairs $\{(W, W^*)\}$ by means of $\mathscr{S}$. Then for each pair from the resulting set, confirmation measure $\mathscr{C}$ is computed using $\mathscr{P}$. Finally, the set of real numbers obtained with $\mathscr{C}$ is transformed by means of $\mathscr{A}$ into a single real number, which represents the quality of the topic interpretability. The scheme for evaluating interpretability in the manner described above is presented in Fig. 1.

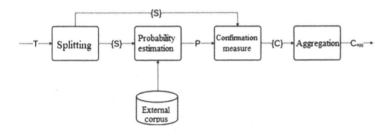

Fig. 1. Scheme of computing an interpretability score

Splitting Method $\mathscr{S}$. To estimate the interpretability of a topic, the set of words representing the topic is divided into pairs, for which their probabilistic "compatibility" is estimated. The most straightforward way of splitting is the simple principle of "every word with every other", S_{one}^{one}. This splitting is used, for example, for the UMI and UMass measures. Further options for splitting include those, in which each word is combined only with each subsequent or with each previous one, S_{suc}^{one} and S_{pre}^{one}. A smarter way of splitting is not only into pairs of single words, but also using subsets of more than one element [6]: we use one versus all other S_{all}^{one}, one versus some subset of other words S_{any}^{one} and two non-intersecting subsets of words S_{any}^{any}.

Probability Estimation Method $\mathscr{P}$. This component determines how the probability of a word is estimated by the external collection of documents. The simplest estimation method, which is used, for example, in the UMass metric, is a method called a "boolean document". The probability of a word is estimated as the number n_w of documents, in which this word occurs, divided by the total number n of documents in the collection. It is worth noting that such estimate of the probability does not take into account the distance between occurrences of words, but only the fact of their appearance in the document.

An alternative approach is the so-called "sliding window". The idea is that a window of fixed size n moves through the external collection of documents. The probability of word w in this case is the number of steps on which w was in the window divided by the total number of steps. In this case, the distance between several words in the text matters, when joint probability is estimated. We will choose from windows of sizes 10, 50, 100 and 200.

Confirmation Measure $\mathscr{C}$. A confirmation measure receives a pair of topic most probable word subsets and uses a probability estimate method considered earlier to calculate how much one subset of the pair is associated with the other. The options, which will be used as elements of the component in the parametric score space, are presented in Table 1.

Table 1. Confirmation measures

Measure	Formula
difference, $\mathscr{C}_d$	$P(W'|W^*) - P(W')$
ratio, $\mathscr{C}_r$	$\frac{P(W',W^*)}{P(W')P(W^*)}$
log-ration, $\mathscr{C}_{lr}$	$\log \frac{P(W',W^*)+\epsilon}{P(W')P(W^*)}$
normalized log-ratio, $\mathscr{C}_{nlr}$	$\frac{m_{lr}(W',W^*)}{-\log(P(W',W^*)+\epsilon)}$
likelihood, $\mathscr{C}_l$	$\frac{P(W'|W^*)}{P(W'|\neg W^*)+\epsilon}$
log-likelihood, $\mathscr{C}_{ll}$	$\log \frac{P(W'|W^*)+\epsilon}{P(W'|\neg W^*)+\epsilon}$
conditional, $\mathscr{C}_c$	$\frac{P(W',W^*)}{P(W^*)}$
logarithmic conditional, $\mathscr{C}_{lc}$	$\log \frac{P(W',W^*)+\epsilon}{P(W^*)}$
Jaccard, $\mathscr{C}_j$	$\frac{P(W',W^*)}{P(W'\vee W^*)}$
logarithmic Jaccard, $\mathscr{C}_{lj}$	$\log \frac{P(W',W^*)+\epsilon}{P(W'\vee W^*)}$
Fitelson [7], $\mathscr{C}_f$	$\frac{P(W'|W^*)-P(W'|\neg W^*)}{P(W'|W^*)+P(W'|\neg W^*)}$

Aggregation Function $\mathscr{A}$. As an aggregation function, we take the arithmetic mean $\mathscr{A}_{am}$, median $\mathscr{A}_{med}$, the geometric mean $\mathscr{A}_{gm}$ and the harmonic mean $\mathscr{A}_{hm}$.

3.2 Regularizer for ARTM

In this Subsection, we describe a new regularizer for ARTM, adding of which will lead to maximizing of a score from the parametric space, maximizing thus the interpretability of the topics.

First, recall that the problem of maximizing the interpretability of topics stands first of all if a person needs to interact directly with a topic model and analyze it. When a person interacts with topics, it is incontinent for a person to consider a topic as a distribution on the whole set of words. Topic models are usually built for large collections of documents, and the dictionary of such collections is so large that a person is not able to process it visually, let alone process a certain number of probability distributions in this dictionary. In this case, the common practice is to present a topic in the form of n of the most probable words. Most often, n is assumed to be 10. Thus, the main idea of the proposed regularizer is to optimize the quality of interpretability of exactly the ten most probable words in the topic.

Let $Top_t = \{w_1, w_2, \ldots, w_{10}\}$ be the ten most probable words of topic t, and $C(u, v)$ be the adjusted confirmation measure for the pair words (u, v), taken from the parametric space of interpretable scores. Then the regularizer, which for each word v from Top_t minimizes the sum of divergences between the distribution of ϕ_{vt} and the confirmation measure for all the remaining words from Top_t, looks like this:

$$R(\Phi) = \tau \sum_{t \in T} \sum_{(u,v) \in Top_t^2} C(u, v) \hat{n}_{ut} \ln \phi_{vt} \to \max,$$

where τ is a regularization coefficient. Further, the resulting regularizer casts the following modified formula for M-step in EM-algorithm:

$$\begin{cases} \phi_{wt} \propto \hat{n}_{wt} + \tau \sum_{v \in Top_t \setminus w} C(w, v) \hat{n}_{vt} & \text{if } w \in Top_t, \\ \phi_{wt} \propto \hat{n}_{wt} & \text{otherwise.} \end{cases}$$

We must note that this regularizer is in general a modification of coherence [12] with specified $C(u, v)$. The confirmation measure must be adjusted with a constant so that its values are to some degree symmetric with respect to zero. That is, for poorly connected words the measure should take negative values, and for well-connected words the value should be positive. Then the presented regularizer can be understood as follows: on each iteration of the algorithm, for each of the ten most probable words of topic t, its relative interpretability to other words from Top_t is estimated; if word w is semantically well-connected with the remaining words $v \in Top_t$ and $C(w, v)$ is positive for the most words, then the probability estimate ϕ_{wt} increases and improves w probability estimation, allowing it to remain in Top_t; if word w is badly related to the rest of $v \in Top_t$, then most values of $C(w, v)$ take negative values, and then the probability estimation ϕ_{wt} decreases, which is likely to cause word w to be excluded from the most probable words of t.

4 Experiment Setup

4.1 Finding Best Interpretability Score in the Parametric Space

Data Labeling. We obtain the topics, on which the score quality is evaluated, using various methods for building topic models, namely PLSA, LDA, and

ARTM. We learn them on a collection of documents representing posts on the blog platform LiveJournal (in Russian). The resulting set counts 1200 topics. Each of the topics is presented with its ten most probable words.

The obtained topics were demonstrated to two assessors. We ask them to estimate each topic, answering two questions. The first question is "Do you understand why these words turned out to be together in this topic?". The second question is "Do you understand what kind of event or phenomenon of life can be discussed in the texts on this topic?". Each answer should be an integer from 0 to 2, where 0 stands for "no", 1 stands for "partly", and 2 stands for "yes". After that each topic was estimated with the mean of two answers.

External Corpus for Learning $\mathscr{P}$**.** In order to learn word probability estimates, we use an external corpus, which is a collection of approximately 1.5 million preprocessed articles of the Russian Wikipedia. First, XML tags and punctuation were removed. Further, all the words were lemmatized by the means of pymorphy2. After that, stop words, numerals, English words, Roman numerals, service parts of speech were removed. Finally, due to the resulting collection was quite large, the index of this collection was built using the Apache Lucene library to improve the speed of work.

Selection Criteria. To assess the quality of the interpretability score, the Spearman rank correlation coefficient between the scores values and the respondents' answers is used. To ensure that the difference between the mean values of the expert estimates is not random, we used Student's t-test.

Experiment Pipeline. We examine each point of the parametric space, with which we estimated each of 1200 topics. We use Java 8 and Palmetto library [18], which implements many elements from which the parametric space components were composed.

4.2   Comparing Topic Models

Document Collection. We use the following document collections: (1) papers presented at conference "Intellectual Data Processing" in various years; (2) articles published in the newspaper "Izvestia" in 1997; (3) text corpus that was labeled within the project OpenCorpora [3].

Each collection was preprocessed in the same way as the external corpus, described in the previous Subsection. Two TMs are built for each of the collections. The first TM is built using such regularizers as the topic rarefaction, blurring of background topics and decorrelation of subject topics. The second TM is build using the very same regularizers and a new regularizer proposed in this work. Each of the six topic models consists of one hundred subject topics and ten background topics.

Topic Model Assessment Criteria. To evaluate how the adding of the new regularizer improves the quality of the interpretability of topics, we invited three assessors who estimated each topic in the way described in the previous Subsection. The most common criterion for the quality of TMs is perplexity, which characterizes the discrepancy between the model $p(w|d)$ for word w observed in documents $d \in D$ and is determined through the log likelihood as follows:

$$\mathscr{P}(D; p) = \exp\left(-\frac{1}{n}\sum_{d \in D}\sum_{w \in d} n_{dw} \ln p(w|d)\right).$$

Also, following the assumption that the topics contain a relatively small number of words from the collection dictionary, and a relatively small number of topics are represented in the documents from the collection, the sparsity of the matrices Φ and Θ is used as an important characteristic of topic models.

Implementation Details. We used BigARTM to implement the TM additive regularization. The source code of this library was supplemented with the regularizer, described in Sect. 3.

5 Results

5.1 Comparison of Interpretability Scores

As a result of the experiment, we found that the highest Spearman coefficient in the interpretability score parametric space was shown by the following score: $(\mathscr{S}_{one}^{one}, \mathscr{P}_{sw(200)}, \mathscr{C}_d, \mathscr{A}_{am},)$ where $\mathscr{S}_{one}^{one}$ is splitting "each word with each other", $\mathscr{P}_{sw(200)}$ is sliding window probability estimation of size 200 words, $\mathscr{C}_d(W', W^*) = P(W'|W^*) - P(W')$, and $\mathscr{A}_{am}$ is the arithmetical mean. It is important to note that in order to use this score in the regularizer presented in Sect. 3, no additional regulation of the confirmation measure by means of some scalars, due to $\mathscr{C}_d$ takes values in range $[-1, 1]$.

We present comparison of the Spearman correlation coefficient (SCC) between the human assessments and all the discussed scores in the Table 2. One can see that the score of the parametric space is superior to all the presented analogs. From this we conclude that the selected interpretability score models human interpretability assessment better than the known scores.

Table 2. Comparison of interpretability scores

Score	SCC	Score	SCC
UCI	0.44538	Context vectors	0.62002
UMass	0.54474	WEC	0.50074
NPMI	0.53320	**ParamSpace**	**0.70330**

As a result, we found a score that maximizes the SCC, which outperforms other scores, and now we can use it for topic modeling.

5.2   Comparison of Topic Model Interpretability and Quality

Comparison Using the Human Assessments. In Table 3 one can see the arithmetic mean of human assessments for each of the built TMs. It is easy to see that addition of the regularizer has significantly improved the interpretability of topics for all collections of documents.

Table 3. Human assessment of the topic models

Collection	Without the regularizer	With the regularizer
ISP	2.357	2.490
Izvestia	2.503	2.863
OpenCorpora	1.950	2.183

The value of Student's t-test was 4.705, which exceeded the value of Student's distribution (2.59) at 299° of freedom and significance level 0.01, which allows rejecting the hypothesis about the equality of mean values.

Comparison Using Topic Model Quality Measures. Figures. 2, 3 and 4 show how the perplexity of TMs has been changing on each step of the EM algorithm. Blue is used by the TMs built using the proposed regularizer, red is used by the TMs built without it. The perplexity of TMs with the regularizer turned out to be noticeably higher, which may indicate that the proposed regularizer worsens the quality of TMs. However, it 2009 in one of the first articles devoted to the interpretability of TMs [4], authors showed that when the value of perplexity is high enough, perplexity and human interpretability assessments are directly dependent. In particular, it was shown that when the perplexity of a TM is reduced, human's interpretability assessments are also reduced. This corresponds to our experiment results described above.

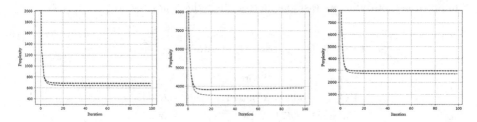

Fig. 2. Perplexity of TMs on IDP corpus   **Fig. 3.** Perplexity of TMs on Izvestia corpus   **Fig. 4.** Perplexity of TMs on OpenCorpora

Figures 5, 6 and 7 show how the sparsity of Φ has been changing during the EM-iterations of the algorithm. It is easy to see that the proposed regularizer somewhat worsens the sparsity of Φ, which looks logical enough given the

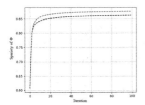

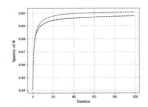

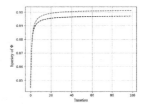

Fig. 5. Sparsity of Θ of TM on IDP corpus

Fig. 6. Sparsity of Θ of TMs on Izvestia corpus

Fig. 7. Sparsity of Φ of TMs on OpenCorpora

structure of the proposed regularizer. However, it caused only a small decrease of the sparseness of Φ (by no more than 2%), which prevents stating that the introduced regularizer significantly worsened the quality of TMs.

The change in the sparsity of Θ during the iterations of the EM algorithm can be traced on Figs. 8, 9 and 10. Interestingly, the introduction of the proposed regularizer somewhat improved the sparsity of the Θ for each of the collections, but not so much as to say that the number of zero elements in the Θ matrix became comparatively large.

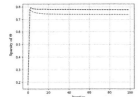

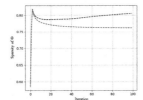

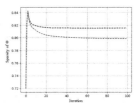

Fig. 8. Sparsity of Θ of TMs on IDP corpus

Fig. 9. Sparsity of Θ of TMs on Izvestia corpus

Fig. 10. Sparsity of Θ of TMs on OpenCorpora

To summarize, the addition of the proposed regularizer did not decreased the quality of TMs, but significantly increased their interpretability.

6   Conclusion

In this paper, we found the best interpretability score in an interpretability score parametric space composed of four components. Basing on this score, we proposed a regularizer for ARTM, which being added is capable of building interpretable topic models. The experiments showed that a topic model with the proposed regularizer significantly outperforms topic models without it having comparable results in term of topic model quality.

As a development of this work, one can consider, for example, the improvement of the semantic similarity of documents belonging to the same topic.

Acknowledgments. Authors would like to thank Anton Belyy and Konstantin Vorontsov for useful conversation. Andrey Mavrin and Andrey Filchenkov were supported by the Government of the Russian Federation (Grant 08-08). Sergei Koltsov was supported by the Basic Research Program at the National Research University Higher School of Economics (HSE).

References

1. Aletras, N., Stevenson, M.: Evaluating topic coherence using distributional semantics. In: Proceedings of the 10th International Conference on Computational Semantics (IWCS 2013)-Long Papers, pp. 13–22 (2013)
2. Blei, D.M., Ng, A.Y., Jordan, M.I.: Latent dirichlet allocation. J. Mach. Learn. Res. **3**(Jan), 993–1022 (2003)
3. Bocharov, V., Bichineva, S., Granovsky, D., Ostapuk, N., Stepanova, M.: Quality assurance tools in the OpenCorpora project (2011)
4. Chang, J., Gerrish, S., Wang, C., Boyd-Graber, J.L., Blei, D.M.: Reading tea leaves: how humans interpret topic models. In: Advances in Neural Information Processing Systems, pp. 288–296 (2009)
5. Daud, A., Li, J., Zhou, L., Muhammad, F.: Knowledge discovery through directed probabilistic topic models: a survey. Front. Comput. Sci. China **4**(2), 280–301 (2010)
6. Douven, I., Meijs, W.: Measuring coherence. Synthese **156**(3), 405–425 (2007)
7. Fitelson, B.: A probabilistic theory of coherence. Analysis **63**(3), 194–199 (2003)
8. Hofmann, T.: Probabilistic latent semantic indexing. In: Proceedings of the 22nd Annual International ACM SIGIR Conference on Research and Development in Information Retrieval, pp. 50–57. ACM (1999)
9. Hong, L., Davison, B.D.: Empirical study of topic modeling in Twitter. In: Proceedings of the First Workshop on Social Media Analytics, SOMA 2010, pp. 80–88. ACM (2010)
10. Islam, A., Inkpen, D.: Second order co-occurrence PMI for determining the semantic similarity of words. In: Proceedings of the International Conference on Language Resources and Evaluation, Genoa, Italy, pp. 1033–1038. Citeseer (2006)
11. Jacobi, C., van Atteveldt, W., Welbers, K.: Quantitative analysis of large amounts of journalistic texts using topic modelling. Digit. Journalism **4**(1), 89–106 (2016)
12. Mimno, D., Wallach, H.M., Talley, E., Leenders, M., McCallum, A.: Optimizing semantic coherence in topic models. In: Proceedings of the Conference on Empirical Methods in Natural Language Processing, pp. 262–272. Association for Computational Linguistics (2011)
13. Newman, D., Karimi, S., Cavedon, L.: External evaluation of topic models. In: 2009 Australasian Document Computing Symposium. Citeseer (2009)
14. Newman, D., Lau, J.H., Grieser, K., Baldwin, T.: Automatic evaluation of topic coherence. In: Human Language Technologies: The 2010 Annual Conference of the North American Chapter of the Association for Computational Linguistics, pp. 100–108. Association for Computational Linguistics (2010)
15. Nikolenko, S.I.: Topic quality metrics based on distributed word representations. In: Proceedings of the 39th International ACM SIGIR Conference on Research and Development in Information Retrieval, pp. 1029–1032. ACM (2016)

16. Papadimitriou, C.H., Tamaki, H., Raghavan, P., Vempala, S.: Latent semantic indexing: a probabilistic analysis. In: Proceedings of the Seventeenth ACM SIGACT-SIGMOD-SIGART Symposium on Principles of Database Systems, pp. 159–168. ACM (1998)
17. Perkio, J., Buntine, W., Perttu, S.: Exploring independent trends in a topic-based search engine. In: 2004 Proceedings of the IEEE/WIC/ACM International Conference on Web Intelligence, WI 2004, pp. 664–668, September 2004
18. Röder, M., Both, A., Hinneburg, A.: Exploring the space of topic coherence measures. In: Proceedings of the Eighth ACM International Conference on Web Search and Data Mining, pp. 399–408. ACM (2015)
19. Rubin, T.N., Chambers, A., Smyth, P., Steyvers, M.: Statistical topic models for multi-label document classification. Mach. Learn. **88**, 157–208 (2012)
20. Steyvers, M., Griffiths, T.: Probabilistic topic models. Handb. Latent Semant. Anal. **427**(7), 424–440 (2007)
21. Vorontsov, K., Frei, O., Apishev, M., Romov, P., Dudarenko, M.: BigARTM: open source library for regularized multimodal topic modeling of large collections. In: Khachay, M.Y., Konstantinova, N., Panchenko, A., Ignatov, D.I., Labunets, V.G. (eds.) AIST 2015. CCIS, vol. 542, pp. 370–381. Springer, Cham (2015). https://doi.org/10.1007/978-3-319-26123-2_36
22. Vorontsov, K., Potapenko, A.: Tutorial on probabilistic topic modeling: additive regularization for stochastic matrix factorization. In: Ignatov, D.I., Khachay, M.Y., Panchenko, A., Konstantinova, N., Yavorskiy, R.E. (eds.) AIST 2014. CCIS, vol. 436, pp. 29–46. Springer, Cham (2014). https://doi.org/10.1007/978-3-319-12580-0_3

Language Resources

Cleaning Up After a Party: Post-processing Thesaurus Crowdsourced Data

Oksana Antropova[1]([✉]), Elena Arslanova[1], Maxim Shaposhnikov[1],
Pavel Braslavski[1,2,3], and Mikhail Mukhin[1]

[1] Ural Federal University, Yekaterinburg, Russia
choksy@mail.ru, contilen@gmail.com, chelseamax@yandex.ru,
{pavel.braslavsky,mikhail.mukhin}@urfu.ru
[2] JetBrains Research, Saint Petersburg, Russia
[3] National Research University Higher School of Economics, Saint Petersburg, Russia

Abstract. The study deals with post-processing of a noisy collection of synsets created using crowdsourcing. First, we cluster long synsets in three different ways. Second, we apply four cluster cleaning techniques based either on word popularity or word embeddings. Evaluation shows that the method based on word embeddings and existing dictionary definitions delivers best results.

Keywords: Crowdsourcing · Thesaurus · Semantic resources

1 Introduction

Thesauri and wordnets are widely used in various natural language processing tasks and applications. There are several approaches to creating wordnets: manual building by professional lexicographers; automatic construction from text corpora and semi-structured sources such as Wiktionary; as well as approaches based on crowdsourcing. In the latter case, non-professional volunteers or paid workers collaboratively construct thesaurus in small steps. Each of the approaches has its pros and cons. For example, crowdsourcing allows quick generation of data at a low cost, but this data is usually noisy and needs post-processing.

In this work, we address the task of analyzing and processing the data generated within the YARN (*Yet Another RussNet*) project [3]. YARN has a web interface[1] allowing virtually everyone to edit existing synsets or create new ones. Manual analysis of these crowdsourced synsets shows that the collection is quite noisy: synsets are duplicated (several non-identical synsets correspond to a concept), may contain irrelevant word entries (synsets often mix synonyms,

---

O. Antropova, E. Arslanova and M. Shaposhnikov contributed equally to the paper.
[1] https://russianword.net/editor.

D. Ustalov et al. (Eds.): AINL 2018, CCIS 930, pp. 133–138, 2018.
https://doi.org/10.1007/978-3-030-01204-5_13

hypernyms and other semantically related words), or are incomplete. Still the synsets created by volunteers have their advantages – they include vocabulary not presented in traditional dictionaries, such as recent borrowings, multi-word expressions, or vulgar words.

In this study, we experiment with several methods aimed at post-processing and cleaning YARN synset collection. First, we cluster long synsets in three different ways. Second, we apply four cleaning techniques based either on word popularity or word embeddings. Evaluation shows that the methods based on embeddings provide better results and can be applied to YARN data.

2   Related Work

Initially, thesauri have been created manually by professional linguists and lexicographers [5]. Crowdsourcing became a viable option for creation and expansion of linguistic resources since its inception in the mid-2000s [6]. For example, [1] describes an experiment on creating a sense inventory using MTurk platform.

YARN project [3] aims at creating an open thesaurus of the Russian language using crowdsourcing. Its user interface [4] allows creating synsets with some guidance from dictionary data. A rather relaxed user action control leads to quite noisy results. A related study [7] describes deduplication of YARN synsets. The authors concluded that three overlapping word entries is an optimal threshold for merging crowdsourced synsets. Cluster cleaning methods using word embeddings employed in the current study are close to synset induction/sense disambiguation methods, see for example an overview in [10]. The difference of the methods herein is that they use dictionary definitions to disambiguate senses.

3   Data

To date, there are 69,796 synsets and 143,508 entries in YARN [2]. 64.3% of synsets contain one or two words, 33.6% – from three to nine, 2.1% – over nine words. In the current study we worked with 23,408 synsets of length from 3 to 9.

Clustering. Following [7], we clustered the collection of synsets in three ways that we denote *GREEDY*, *TRIPLES*, and *BABENKO*. *GREEDY* is a variant of single linkage clustering – synsets sharing three words are clustered, which results in 16,694 clusters. In case of *TRIPLES* each word triple occurring in the initial synsets defines a cluster. It is a variant of soft clustering: synsets can be assigned to multiple clusters. This process generates 20,966 clusters. The third option makes use of a machine-readable dictionary of Russian synonyms.[3] The dictionary contains 29,194 entries organized into 7,538 synsets. For each YARN synset in the initial collection we searched for the closest dictionary synset in terms of Jaccard coefficient and clustered synsets belonging to the same

---

[2] https://russianword.net/data.
[3] Babenko, L.G.: The thesaurus dictionary of the Russian language synonyms, 2008.

BABENKO synset together. The rationale behind this methods is to enrich dictionary synsets with crowdsourced multi-word expressions, recent borrowings, etc. *BABENKO* clustering resulted in 9,323 clusters (YARN synsets not linked to *BABENKO* synsets become single-synset clusters).

Collection of Definitions. We also use dictionary definitions to re-organize synset clusters. The majority of definitions employed in the study come from Wiktionary[4]. Missing definitions were collected from dictionaries by Efremova, Ozhegov, Ushakov, and Babenko.[5] In total, the collection comprises 187,003 unique definitions for 132,485 word entries.

Gold Synsets. In order to tune parameters of the methods and evaluate them, we manually created a small collection of 'gold synsets'. We started with 1,140 noisy synsets. First, we manually clustered them in such a way that synsets in a cluster describe the same concept, which resulted in 139 clusters. Second, a gold synset was created for each cluster by removing duplicate and irrelevant words. The gold synsets were randomly split into training set (39) used for parameter tuning and test set (100).

4 Methods

4.1 Words and Synsets Weighing

Two methods exploiting redundancy produced by the crowd were developed for synset cluster cleaning: *WordWeights* and *SynsetRanks*.

WordWeights method is based on a simple idea: the more synsets in the cluster contain the word the more likely it is actually relevant to the concept. Thus, for every word in a cluster, we calculate its weight as a share of initial synsets it occurs in. If the word weight exceeds a threshold, the word is added to the 'pure' synset representing the cluster. The threshold is optimized on the training set.

SynsetRanks aims at estimating the quality of the initial synset as a whole and then compiling a 'clean' synset from good ones. First, synsets in a cluster are ranked based on their weights. Synset weight is calculated as average of its *WordWeights*. Synsets below a threshold are discarded. The remaining synsets are merged incrementally top-down if their similarity exceeds the second threshold. Both parameters of the routine are tuned on the training set.

4.2 Cleaning with Embeddings

The second group of methods is based on word embeddings and collection of dictionary definitions. We employed pre-trained *fasttest* 300-dimensional vectors

---

[4] https://ru.wiktionary.org/.
[5] An overview of dictionary data available for Russian can be found in [8].

from RusVectōrēs project [9][6]. *Fasttext* vector representation combines word-level and character n-grams embeddings, which is helpful in case of highly inflectional Russian language and partly mitigates out-of-vocabulary problem [2]. Gensim[7] library was used to query the model and calculate cosine word similarities.

Word2vec[8] uses vector representations of words (in case of multiwords a sum of constituents' vectors is used). First, we calculate all pairwise similarities of words in the cluster. Second, we rank words according to the average similarity to all cluster members. Thus, more central words are ranked higher. Then we incrementally build new synsets from the top of the list by adding words if their average similarity to the items already in synset exceeds a threshold.

Def2vec model represents word senses as an average of vectors constituting their definitions. By this approach, each word is represented by a *set* of vectors, each reflecting one of its senses. After initial pre-ordering of the words as in the *Word2vec* model the extension of the classic agglomerative clustering algorithm is performed. The main advantage of the approach is that we account for polysemy; moreover, the newly built synsets are delivered with definitions.

5   Results and Discussion

For every 'golden' cluster we found the closest newly obtained cluster by Jaccard similarity. Then we applied a cleaning method with all the possible parameters to every cluster aligned with the training set and chose the parameters that provided the best results. After that we applied all the methods with the optimized parameters to the test set and evaluated the results, see Table 1.

Table 1. Evaluation results. J – Jaccard coefficient, P – precision, R – recall, $F1$ – F1-score.

	GREEDY				TRIPLES				BABENKO			
Method	J	P	R	$F1$	J	P	R	$F1$	J	P	R	$F1$
WordWeights	0.42	0.58	0.63	0.54	0.53	0.62	0.84	0.67	0.48	0.76	0.60	0.60
SynsetRanks	0.40	0.59	0.53	0.51	0.55	0.68	0.80	0.68	0.46	0.70	0.58	0.58
Word2vec	0.51	0.70	0.69	0.65	0.55	0.73	0.74	0.69	0.49	0.68	0.66	0.63
Def2vec	0.45	0.77	0.57	0.60	0.52	0.81	0.63	0.66	0.45	0.68	0.64	0.60

As long as the proposed evaluation method estimates only the clusters aligned with the 'gold' clusters, the results shown in the Table 1 are overestimated to

---

[6] http://rusvectores.org/ru/models/.

[7] https://radimrehurek.com/gensim/.

[8] We use *word2vec* as a name of a general approach to word embeddings and to contrast it to the latter method that works with definitions.

some extent. Especially strongly it affects the results of *TRIPLES*, because it creates many small clusters (43% of aligned clusters consist of one or two synsets). Simple alignment of completely unprocessed YARN synsets with the 'gold' synsets provides *Jaccard* = 0.71 and *F1* = 0.81, so it explains why *TRIPLES* delivers the highest values. Nonetheless, we would not recommend using this method, because it is overrated and leaves most of the duplicates unclustered.

GREEDY tends to mix different senses of polysemic words. For example, it unites different senses of the word **край**: *land, country, territory, side* and *brim*. Success of *WordWeights* and *SynsetRanks* in such cases depends on what concept dominates in YARN (usually this is the most frequent sense). *Word2vec* and *Def2vec* manage well as long as relevant contexts and definitions are found. *BABENKO* clusters are noticeably cleaner and usually contain more synsets, which work best for redundancy-based methods. *Word2vec* and *Def2vec*, in opposite, do not succeed with this clustering, because *BABENKO* clusters usually contain closely related words that have similar definitions and occur in similar contexts. Table 2 illustrates these considerations with the synsets obtained for the concept *state, country*.

Table 2. Results aligned with the 'gold' sysnset {*держава*[1], *государство*[2], *страна*[3]}

Method	GREEDY	BABENKO
Word Weights	*область*[4], *сторона*[5], *территория*[6], *регион*[7], *страна*[3], *государство*[2], *местность*[8], *край*[9]	*держава*[1], *государство*[2], *страна*[3]
Synset Ranks	*область*[4], *местность*[8], *край*[9], *страна*[3], *округа*[10]	*держава*[1], *государство*[2], *страна*[3]
Word2vec	*держава*[1], *государство*[2], *царство*[11], *княжество*[12], *империя*[13], *страна*[3]	*государственная власть*[14], *содружество*[15], *государство*[2], *империя*[13], *страна*[3], *нация*[16], *государственное управление*[17], *держава*[1]
Def2vec	*королевство*[18], *держава*[1]	*государство*[2], *империя*[13], *царство*[11], *нация*[16], *держава*[1], *страна*[3], *королевство*[18], *содружество*[15], *территория*[6]

[1] *(literary)* state, [2] state, [3] country, [4] province, [5] side, [6] territory, [7] region, [8] area, [9] land, [10] neighbourhood, [11] realm, [12] princedom, [13] empire, [14] government, [15] commonwealth, [16] nation, [17] state authorities, [18] kingdom

Despite the fact that redundancy methods demonstrate relatively good results in case of *BABENKO* clustering, in fact they produce only synsets related to the most general concepts (because such words usually dominate), and discard all the data related to more specific concepts. *Word2vec* and *Def2vec* keep all words from initial YARN synsets intact. *Def2vec* delivers higher precision and additionally provides definitions for newly created synsets. *Word2vec* works better if there are no relevant meanings in the dictionary. For example, it generates a correct synset {*штрих, корректор*} for the concept *correction fluid*, whereas *Def2vec* splits the pair according to their dictionary meanings: *stroke* for *штрих* and *corrector* (profession) for *корректор*.

6   Conclusion

The quality of crowdsoursed YARN synsets varies greatly. Most of them mixes two and more similar concepts and are incomplete at the same time. Nonetheless, as a whole they cover significantly more vocabulary than traditional synonym dictionaries. The proposed methods of post-processing allow to improve average quality of synsets, but they can hardly distinguish synonyms from hyponyms/hypernyms and co-hyponyms. Thus, the study confirms that crowdsourced projects demand well-thought user action control and organization.

Word2vec provides best recall among the examined methods and can be recommended if followed by manual editing by a qualified lexicographer. Otherwise, *Def2vec* delivering highest precision is quite a practical option. We plan to apply *GREEDY* clustering and *Def2vec* cleaning to YARN data.

Acknowledgments. PB was supported by RFH grant #16-04-12019, OA was supported by RFBR according to the research project No. 18-312-00129.

References

1. Biemann, C.: Creating a system for lexical substitutions from scratch using crowdsourcing. Lang. Resour. Eval. **47**(1), 97–122 (2013)
2. Bojanowski, P., Grave, E., Joulin, A., Mikolov, T.: Enriching word vectors with subword information. arXiv preprint arXiv:1607.04606 (2016)
3. Braslavski, P., Ustalov, D., Mukhin, M., Kiselev, Y.: YARN: spinning-in-progress. In: GWC, pp. 58–65 (2016)
4. Braslavski, P., Ustalov, D., Mukhin, M.: A spinning wheel for YARN: user interface for a crowdsourced thesaurus. In: EACL (demo), pp. 101–104 (2014)
5. Fellbaum, C.: Wordnet: An Electronic Database. MIT Press, Cambridge (1998)
6. Gurevych, I., Kim, J. (eds.): The People's Web Meets NLP. Springer, Heidelberg (2013). https://doi.org/10.1007/978-3-642-35085-6
7. Kiselev, Y., Ustalov, D., Porshnev, S.: Eliminating fuzzy duplicates in crowdsourced lexical resources. In: GWC, pp. 161–167 (2016)
8. Kiselev, Y., et al.: Russian lexicographic landscape: a tale of 12 dictionaries. In: Dialogue, pp. 254–271 (2015)
9. Kutuzov, A., Kuzmenko, E.: Webvectors: a toolkit for building web interfaces for vector semantic models. In: AIST, pp. 155–161 (2017)
10. Ustalov, D., Panchenko, A., Biemann, C.: Watset: automatic induction of synsets from a graph of synonyms. In: ACL, pp. 1579–1590 (2017)

A Comparative Study of Publicly Available Russian Sentiment Lexicons

Evgeny Kotelnikov$^{(\boxtimes)}$, Tatiana Peskisheva, Anastasia Kotelnikova,
and Elena Razova

Vyatka State University, Kirov, Russia
kotelnikov.ev@gmail.com, peskisheva.t@mail.ru,
kotelnikova.av@gmail.com, razova.ev@gmail.com

Abstract. Sentiment lexicons play an important role in the systems of sentiment analysis and opinion mining. The article takes a look into eight publicly available Russian sentiment lexicons of today. A joint analysis of these lexicons was done by finding unions and intersections of the lexicons and also analysing the distribution of parts of speech. In order to study the quality of the lexicons, a sentiment classification is made based on the SVM and the TF-IDF model. Text corpora from reviews of works of art (books and movies), organizations (banks and hotels) and goods (kitchen appliances) are made for this purpose. Lexicons are compared in terms of their classification quality, and also on the basis of a linear regression model that reflects the dependence of their F1-measure on their TF-IDF model size. The resulting union lexicon most fully reflects the sentiment lexica of the present day Russian language and can be used both in scientific research and in applied sentiment analysis systems.

Keywords: Sentiment lexicons · Sentiment analysis · Opinion mining

1 Introduction

Currently there are three main approaches to text opinion mining and sentiment analysis – machine learning, lexicon-based approach and hybrid approach [8, 23]. Machine learning techniques require well labelled training data and spend considerable time on the training procedure [6]. The lexicon-based methods do not have these drawbacks, but their analysis accuracy is often not high enough [7]. Hybrid systems combine different approaches [2].

Sentiment lexicons are a key element in the two latter approaches. The quality of the sentiment analysis in this case will be determined by the quality of such lexicons. Many papers have been dedicated to building lexicons for sentiment analysis [20, 21, 24], but the attention paid to the problem of existing lexicons research remains insufficient.

Another important problem is choosing the most effective sentiment lexicon, which provides the best performance for sentiment analysis. For example, at the moment there are at least 8 sentiment lexicons for the Russian language (see Sect. 3). The choice of the optimal lexicon among so many lexicons turns out to be non-trivial.

D. Ustalov et al. (Eds.): AINL 2018, CCIS 930, pp. 139–151, 2018.
https://doi.org/10.1007/978-3-030-01204-5_14

The contribution of this paper is as follows: (1) a study of publicly available Russian sentiment lexicons (Sect. 3); (2) undertaking a joint analysis of them (Sect. 4); (3) an evaluation of the quality of the sentiment analysis with the use of these lexicons, as well as with the union dictionary based on them (Sect. 5).

2 Related Work

There have been a few studies of existing sentiment lexicons in themselves without reference to specific problems of opinion mining in recent years [4, 11, 17, 19, 24].

Chen and Skiena [4] automatically build sentiment lexicons for 136 languages (including Russian) and compare them with existing ones for six languages (Arabic, Chinese, English, German, Italian, Japanese). The authors evaluate the accuracy and coverage of the sentiment lexicons. Accuracy is the proportion of words of the same polarity between the built and the published lexicons (from 56% to 97%) and coverage is a fraction of the built lexicons overlap with published lexicons (from 12% to 72%).

Potts [19] analyzes five English lexicons: Bing Liu's Opinion Lexicon [13], SentiWordNet [1], MPQA [26], General Inquirer [22], and Linguistic Inquiry and Word Counts (LIWC) [12]. For each pair of lexicons their disagreement level is evaluated (from 0.5% to 27%). Disagreement level shows how often they explicitly disagree with each other in that they supply opposite polarity values for a given word.

In [24] the English-language lexicon created by authors manually, as well as other sentiment lexicons, including SentiWordNet and MPQA, are rated by people using the Amazon Mechanical Turk service.

The articles closest to our work are [11, 17]. Ohana et al. [17] compare four English sentiment lexicons for sentiment classification, including SentiWordNet, MPQA and General Inquirer. The General Inquirer lexicon is used as the baseline dictionary and the degree of agreement with it for the remaining lexicons is calculated (from 31.4% to 85.9%). Also the quality of the sentiment analyses is compared for different lexicons and for ensembles of classifiers built on their basis. It turns out that the ensembles show a higher quality of classification. In our work we don't use the ensembles of classifiers, but a single classifier on the basis of all lexicons combination. In addition, all available Russian sentiment lexicons are researched.

Kotelnikov et al. [11] analyze Russian dictionaries, created by four annotators in five domains. The inter-annotator agreement, the distribution of parts of speech in the lexicons for different domains, the degree of agreement between manual and automatic lexicons (from 6% to 33%) are given. Also, the quality of the sentiment analysis based on manual, automatic and published lexicons is compared. In our work all available sentiment lexicons are examined, unlike [11], where only three existing lexicons are analyzed. We also build and analyze a union lexicon and carry out a classification based on it.

Kiselev et al. [9] explore 12 existing Russian lexical-semantic resources (printed explanatory dictionaries, dictionaries of synonyms, electronic thesauri). The degree of overlap between the dictionaries, the number of unique words in the dictionaries, the degree of coverage of the corpora, the number of neologisms, the analysis of synonyms and definitions are given. Although the paper doesn't cover sentiment lexicons, analytical methods used are close to our research.

3 Russian Sentiment Lexicons

Currently there are at least 8 Russian sentiment lexicons which are publicly available (see Table 1).

1. ProductSentiRus
The ProductSentiRus lexicon was developed by I. Chetviorkin and N. Loukachevitch for the product meta-domain (movies, books, games, digital cameras, mobile phones) [5]. The set of statistical and linguistic features of evaluation words and machine learning algorithms were used. As a result, a general sentiment lexicon from five domains with the quantity of 5000 words was obtained. The words in the lexicon are ordered by the probability of their sentiment polarity, but they are not divided into positive and negative.

2. Blinov et al.'s Lexicon
Blinov et al. [3] manually selected a list of 969 most positive and 1,138 most negative words from the ProductSentiRus lexicon, and then automatically expanded the list with synonyms and antonyms from Russian Wiktionary.

3. EmoLex
The EmoLex lexicon (NRC Emotion Lexicon) was compiled by Mohammad and Turney with the use of crowdsourcing [16]. The lexicon contains 14,182 words, correlated with positive and negative sentiment, as well as with such emotions as anger, anticipation, disgust, fear, joy, sadness, surprise and trust. The lexicon was translated to more than 100 languages (including Russian) using Google Translate. We selected words with only positive or only negative sentiment for our research (excluding collocations).

3. Chen-Skiena's Lexicon
Chen and Skiena [4] built sentiment lexicons for 136 languages. On the basis of Wiktionary, Google Translate, transliteration links and WordNet, a knowledge graph was built, connecting words in different languages. After that from seed English sentiment lexicon they construct sentiment lexicons for each component language using graph propagation.

5. Tutubalina's Lexicon
Tutubalina in her thesis [25] manually created a lexicon on the basis of strictly positive and negative users' reviews about cars (only sections, containing advantages and disadvantages in the reviews were taken into account). The lexicon was expanded by adding synonyms.

6. Kotelnikov et al.'s Lexicon
Kotelnikov et al. [11] first automatically selected 10,000 candidate words from each of five domains (user reviews of restaurants, cars, movies, books and digital cameras) to obtain sentiment lexicon. Four annotators assessed each word as positive, negative,

neutral or contradictory, then two sentiment lexicons (including only positive and negative words) united by the domains were created: the first one included words about the sentiment of which three annotators out of four agreed (we denote it as "Kotelnikov et al.'s lexicon (large)"), the second one contained the words about the sentiment of which all annotators agreed ("Kotelnikov et al.'s lexicon (small)").

7. LinisCrowd

Koltsova et al. [10] created their lexicon using crowdsourcing. At first they selected 7,546 words based on a list of high-frequency adjectives, the ProductSentiRus lexicon, the explanatory dictionary, and the translation of the English sentiment lexicon. Then, each word was rated from -2 up to 2 at least by three annotators. We considered words as positive and as negative if they received the majority of ratings of the corresponding sentiment.

8. RuSentiLex

Loukachevitch and Levchik [15] created the RuSentiLex lexicon, in which for each word the sentiment (positive, negative, neutral) and the source (opinion, fact, feeling) are indicated. At first lists of sentiment words based on the RuThes thesaurus [14], existing sentiment lexicons, news articles and Twitter were generated, then the linguists analyzed the resulting lists to create the final lexicon.

In our work we used the version of the RuSentiLex of 2017 and only words with positive and negative sentiment (collocations were not used). We tested two versions – large (we denote it as "RuSentiLex (large)"), including all positive and negative words, and small ("RuSentiLex (small)"), including positive and negative words, for which source is equal to opinion.

Thus, taking into account the two versions of Kotelnikov et al.'s lexicon and RuSentiLex, 10 lexicons were explored in the paper. In Table 1 the characteristics of the considered lexicons are given: the sizes of positive and negative word sets, the sizes of the unions and intersections of these sets, the approach used to create the lexicon, and the year of creation. For five lexicons the intersections of positive and negative word sets are non-empty. Apparently, in the lexicons of Blinov et al. and Tutubalina it happened because of the automatic expansion of the original word lists with synonyms, and in the EmoLex – due to automatic translation. In the Kotelnikov et al.'s lexicon the same word can be positive for one domain and negative for another domain, for example, *непредсказуемый сюжет – непредсказуемые отказы (unpredictable plot – unpredictable failures)*. In the RuSentiLex lexicon the same words can also have different meaning and sentiment, it is shown in the reference to the article of the thesaurus RuThes, for example, *легкий (покладистый) – легкий (поверхностный) (easy (flexible) – easy (superficial))*.

The average number of negative sentiment words is almost twice as high as the number of positive words – negative vocabulary is more diverse.

Table 1. Characteristics of Russian sentiment lexicons

Lexicons	pos	neg	union	intersection	method	year
ProductSentiRus			5,000	0	auto	2012
Blinov et al.'s lexicon	1,864	2,145	3,839	170	auto + manual	2013
EmoLex	1,974	2,575	4,412	137	manual	2013
Chen-Skiena's lexicon	1,246	1,630	2,876	0	auto	2014
Tutubalina's lexicon	1,078	1,458	2,509	27	manual	2016
Kotelnikov et al.'s lexicon (large)	1,046	2,209	3,245	10	auto + manual	2016
Kotelnikov et al.'s lexicon (small)	387	724	1,110	1	auto + manual	2016
LinisCrowd	566	1,940	2,506	0	auto + manual	2016
RuSentiLex (large)	2,794	7,882	10,543	133	auto + manual	2017
RuSentiLex (small)	2,341	4,886	7,153	74	auto + manual	2017
Average	**1,478**	**2,828**	**4,319**	**55**		

4 Analysis of Lexicons

4.1 Union and Intersections

The union of all lexicons includes 20,401 words: 7,915 positive (38.8%)[1], 11,768 negative (57.7%), and 718 words that could not be uniquely determined for the sentiment (3.5%). The intersection of all lexicons (denoted as *Intersection10*) contains 35 words: 26 positive and 9 negative (Table 2). In the set of common words for all lexicons there are 33 adjectives and two nouns (*красота* (beauty) and *преимущество* (advantage)). It is interesting that such words as *хороший* (good) and *плохой*[2] (bad) did not fall into this set. To exclude such outliers, we also found a set of words which appear in at least 9 lexicons out of 10 – this is the union of all the intersections of 9 lexicons (*Intersection9*). This set includes 143 words: 75 positive (70 adjectives and 5 nouns) and 68 negative (56 adjectives, 10 nouns, 1 verb and 1 adverb).

---

[1] The sentiment of the word was determined by voting on the majority of lexicons to which it belongs.

[2] *Хороший* (good) is absent in EmoLex (the word *good* was translated as *хорошо*), *плохой* (bad) is absent in Kotelnikov et al.'s lexicon (small) (one of the annotators attributed *bad* to neutral words for several domains).

Table 2. Intersection of all sentiment lexicons

Positive words		Negative words
благоприятный (favorable), *великолепный (excellent),* *волшебный (magical),* *достойный (worthy),* *замечательный (wonderful),* *идеальный (ideal),* *красивый (beautiful),* *красота (beauty),* *легендарный (legendary),* *надежный (reliable),* *недорогой (inexpensive),* *нежный (delicate),* *превосходный (perfect),*	*преимущество (advantage),* *прекрасный (beautiful),* *привлекательный (attractive),* *приличный (decent),* *приятный (pleasant),* *роскошный (luxurious),* *удивительный (amazing),* *удобный (comfortable),* *ценный (valuable),* *чудесный (marvelous),* *энергичный (energetic),* *эффективный (effective),* *яркий (bright)*	*бессмысленный (meaningless),* *глупый (stupid),* *грязный (dirty),* *неприятный (unpleasant),* *неудачный (unsuccessful),* *печальный (sad),* *трудный (difficult),* *тупой (dumb),* *ужасный (horrible)*

Table 3 shows the ratio of cardinalities of pairwise lexicons intersections to pairwise lexicons unions (in percent) as a heatmap.

Table 3. Ratios of cardinalities of pairwise lexicons intersections to pairwise lexicons unions (%)

Lexicons	ProductSentiRus	Blinov et al.'s lexicon	EmoLex	Chen-Skiena's lexicon	Tutubalina's lexicon	Kotelnikov et al. (large)	Kotelnikov et al. (small)	LinisCrowd	RuSentiLex (large)
Blinov et al.'s lexicon	33.4								
EmoLex	10.4	11.1							
Chen-Skiena's lexicon	11.2	11.0	18.2						
Tutubalina's lexicon	13.8	21.4	9.3	6.7					
Kotelnikov et al. (large)	17.2	21.2	9.1	8.0	15.2				
Kotelnikov et al. (small)	9.3	12.9	5.6	5.1	11.8	34.2			
LinisCrowd	12.4	15.0	13.1	9.7	15.0	15.8	12.2		
RuSentiLex (large)	10.8	13.1	12.2	8.2	11.3	12.9	6.0	16.7	
RuSentiLex (small)	11.8	14.5	11.1	7.5	14.3	14.2	7.4	17.1	67.8

The high values of the ratios between the RuSentiLex (small) and RuSentiLex (large), and also Kotelnikov et al.'s lexicon (small) and Kotelnikov et al.'s lexicon (large) are due to the fact that small versions are subsets of large versions. Blinov et al.'s lexicon is based on the ProductSentiRus lexicon (33.4%). There is a great similarity between the three lexicons: Blinov et al.'s lexicon, Kotelnikov et al.'s lexicon (large) and Tutubalina's lexicon – an average of 19.3%.

In general, the coincidence of the words between the lexicons is very small (on average 13.7% without taking into account the lexicons-subsets RuSentiLex and Kotelnikov et al.'s lexicon). The translated lexicons (EmoLex and Chen-Skiena's lexicon) have a smaller intersection with the rest of the lexicons than the average (10.0%), and at the same time are relatively similar to each other (18.2%).

4.2 Parts of Speech

Table 4 shows the distribution of parts of speech in separate lexicons and in the union lexicon. Parts of speech were obtained with the help of mystem parser[3]. Words that mystem parser did not recognize were placed in the "Unknown" column. As a rule, these are either slang words ("*улетный*" (*ulyotnyj, flying away*), "*тупизм*" (*tupizm, stupidity*)), or words with misspellings (*lutshij* instead of *luchshij* (*better*), *brakovanyj* instead of *brakovannyj* (*defective*))[4].

Table 4. Distribution of parts of speech (%)

Lexicons	Nouns	Verbs	Adject.	Adv.	Others	Unkn.	Sum
ProductSentiRus	14.0%	28.4%	39.6%	11.9%	1.1%	5.0%	100%
Blinov et al.'s lexicon	16.9%	22.5%	46.1%	9.7%	0.4%	4.4%	100%
EmoLex	55.0%	16.8%	25.2%	2.3%	0.4%	0.3%	100%
Chen-Skiena's lexicon	48.9%	22.7%	20.0%	7.8%	0.7%	0.0%	100%
Tutubalina's lexicon	6.5%	7.6%	73.9%	2.4%	0.0%	9.6%	100%
Kotelnikov et al.'s lexicon (large)	26.6%	26.0%	34.1%	13.2%	0.1%	0.0%	100%
Kotelnikov et al.'s lexicon (small)	26.7%	12.9%	44.3%	16.1%	0.0%	0.0%	100%
LinisCrowd	38.0%	18.2%	43.0%	0.6%	0.2%	0.0%	100%
RuSentiLex (large)	41.4%	23.3%	24.5%	0.4%	0.0%	10.4%	100%
RuSentiLex (small)	41.8%	17.9%	30.1%	0.5%	0.0%	9.6%	100%
Union lexicon (20,401 words)	36.1%	25.7%	24.9%	5.1%	0.5%	7.8%	100%

---

[3] https://tech.yandex.ru/mystem/.

[4] Mystem hypothesizes a part of speech (usually true for slang words and often erroneous for words with misspellings), but since this information is not fully reliable, we decided to add the column "Unknown".

Several lexicons prefer some parts of speech to others, for example, there are many adjectives in Tutubalina's lexicon, Blinov et al.'s lexicon and in Kotelnikov et al.'s lexicon (small); in translated lexicons, as well as in RuSentiLex, the proportion of nouns is high. In Kotelnikov et al.'s lexicon (large) and LinisCrowd the ratio of nouns and adjectives is relatively balanced. It is interesting to note the high proportion of adverbs in ProductSentiRus and Kotelnikov et al.'s lexicon: these are such words as *безупречно (bezuprechno, flawlessly), солидно (solidno, solidly), гадко (gadko, disgustingly), скучно (skuchno, drearily).*

There is a high proportion of slang words and words with typos in ProductSentiRus, Blinov et al.'s lexicon, Tutubalina's lexicon, RuSentiLex. It should be noted that in RuSentiLex there are almost no words with misspellings, apparently due to the processing of the lexicon by linguists.

Nouns (36.1%) prevail in the union lexicon, and there is approximately the same number of verbs and adjectives (about 25%) and 5% of adverbs. Compared with the *Intersection10* and *Intersection9* lexicons (see Sect. 4.1), containing approximately 90% of adjectives, their share decreased significantly. Thus, the core of the sentiment lexicon (i.e., the vocabulary on which all or almost all lexicons agree) is made up by adjectives, but as the lexicon expands, the nouns begin to predominate.

5 Sentiment Classification

5.1 Text Corpora

To study the quality of sentiment classification on the basis of sentiment lexicons, text corpora of reviews of books and movies, organizations (banks and hotels) and goods (kitchen appliances) were created[5]. From several tens to several hundreds of thousands of reviews (see Table 5) were collected, after what 10,000 reviews were selected for

Table 5. Characteristics of text corpora

Domain	No. reviews before selection	Rating scales	No. reviews after selection	No. pos	No. neg
Books	216,358	[0.5... 5] with step 0.5	10,000	5,000	5,000
Movies	103,668	{bad = 1, neu = 3, good = 5}	10,000	5,000	5,000
Banks	45,189	{1, 2, 3, 4, 5}	10,000	5,000	5,000
Hotels	109,999	[0... 5] with step 0.1	10,000	5,000	5,000
Kitchen	106,862	{2, 3, 4, 5}	10,000	5,000	5,000
Total	582,076		50,000	25,000	25,000

---

[5] Sources of the reviews: www.kinopoisk.ru, tophotels.ru, www.banki.ru, www.e-katalog.ru.

each domain. For this selection the reviews were sorted by length, 5% of the shortest and longest reviews were rejected, then the reviews were randomly selected. The original rating scales were transformed into binary scale (positive – negative) according to the following scheme: $[0...2.5] \rightarrow neg$, $[4...5] \rightarrow pos$ (the values between 2.5 and 4 are discard). Morphological analysis of all reviews was performed using mystem parser.

5.2 Results and Discussion

We studied the quality of the sentiment classification using different sentiment lexicons, as well as union lexicon. As a classifier the SVM from the library scikit-learn [18] was used. We have also tried other classifiers, e.g. Naïve Bayes classifier and k-Nearest Neighbors, but they have shown worse performance. The classifier parameters were selected on the basis of 5–fold cross-validation independently for each domain using special validation corpora, containing 2,000 reviews each, selected according to the same procedure as the main corpora. The following parameter values were examined: {*kernel*: *linear, rbf, polynomial*}; {C: 0.1, 1, 10, 100}, {*gamma*: 0, 0.1, 0.01}. As a result, linear kernel and $C = 1$ turned out to be optimal for all domains.

For each domain a TF-IDF text representation model (from scikit-learn) was created which included words from only given sentiment lexicon. The number of words in the models is indicated in Table 6. Also the TF-IDF model was investigated on the basis of the full corpus dictionary for each domain.

Table 6. TF-IDF models (number of words)

Lexicon	Books	Movies	Banks	Hotels	Kitchen	Average
ProductSentiRus	4,009	4,073	3,188	3,579	3,112	3,592
Blinov et al.'s lexicon	2,918	3,021	2,128	2,502	1,990	2,512
EmoLex	3,476	3,559	2,314	2,827	1,893	2,814
Chen-Skiena's lexicon	2,199	2,249	1,789	1,984	1,568	1,958
Tutubalina's lexicon	1,641	1,694	1,114	1,342	966	1,351
Kotelnikov et al.'s (large)	2,679	2,818	1,861	2,458	1,920	2,347
Kotelnikov et al.'s (small)	981	1,016	709	923	673	860
LinisCrowd	2,269	2,305	1,474	1,780	1,113	1,788
RuSentiLex (large)	5,910	6,198	3,204	4,059	2,363	4,347
RuSentiLex (small)	4,281	4,470	2,399	2,881	1,769	3,160
Union lexicon	12,127	12,674	7,459	9,423	6,518	9,640
Full dictionary of corpus	37,310	38,349	19,304	28,115	15,736	27,763

Table 7 presents the results of the sentiment classification (F1-measure) for all lexicons and domains, and also on average for domains.

Table 7. Results of classification, F1-meaure

Lexicon	Books	Movies	Banks	Hotels	Kitchen	Average
ProductSentiRus	**0.812**	**0.868**	**0.919**	**0.938**	**0.913**	**0.890**
Blinov et al.'s lexicon	0.799	0.858	0.899	0.931	0.898	0.877
EmoLex	0.765	0.817	0.885	0.919	0.865	0.850
Chen-Skiena's lexicon	0.756	0.820	0.888	0.915	0.873	0.851
Tutubalina's lexicon	0.719	0.793	0.847	0.899	0.850	0.821
Kotelnikov et al.'s (large)	0.775	0.819	0.891	0.925	0.887	0.860
Kotelnikov et al.'s (small)	0.739	0.796	0.852	0.908	0.824	0.824
LinisCrowd	0.722	0.792	0.832	0.905	0.830	0.816
RuSentiLex (large)	0.763	0.823	0.867	0.920	0.859	0.846
RuSentiLex (small)	0.739	0.810	0.840	0.905	0.834	0.826
Union lexicon	0.834	0.885	0.934	0.949	0.925	0.905
Full dictionary of corpus	**0.847**	**0.891**	**0.942**	**0.955**	**0.932**	**0.913**

Among the sentiment lexicons the best results are demonstrated by ProductSentiRus (on average of 0.89). It surpasses Blinov et al.'s lexicon (second place) by 0.013, and Kotelnikov et al.'s large lexicon (third place) by 0.03. The worst results are shown by LinisCrowd, Tutubalina's lexicon and Kotelnikov et al.'s small lexicon, lagging behind ProductSentiRus by 0.074, 0.069 and 0.066 respectively.

The results of the union lexicon for all domains outperform ProductSentiRus results by 0.015 on average. In turn, TF-IDF models, built on the basis of a full dictionary of corpus, turn out to be on average 0.008 better than those based on the union lexicon.

Based on the average size of the TF-IDF models and the average quality of the classification (F1-measure) for each sentiment lexicon (see last columns in Tables 6 and 7), we calculated the Pearson's correlation coefficient between quality and the size of the TF-IDF model. It turned out to be equal to 0.5, which indicates the presence of dependence.

However, for some lexicons this dependence is more expressed than on average, and for some – weaker. Figure 1 shows the dependence of the F1-measure on the number of words in the TF-IDF model, as well as the linear regression line.

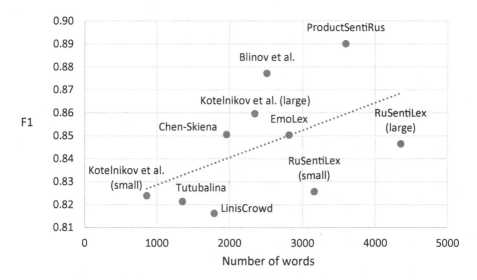

Fig. 1. A dependence of the F1-measure on the number of words in the TD-IDF model.

The lexicons located above the linear regression line (ProductSentiRus, Blinov et al.'s lexicon, Kotelnikov et al.'s large lexicon and Chen-Skiena's lexicon) demonstrate a higher quality of classification than it could be expected in accordance with the size of TF-IDF models, and vice versa, lexicons located below this line (both RuSentiLex lexicons, Tutubalina's lexicon, LinisCrowd and union lexicon[6]) show a lower quality of classification than expected. Kotelnikov et al.'s small lexicon and EmoLex roughly correspond to the expected quality of the classification. It should be noted that the full dictionary of corpus is located much lower than the linear regression line.

6   Conclusion

Thus, publicly available Russian-language sentiment lexicons, created from 2012 to 2017, differ significantly: in size (from 1,110 to 10,543 words), in lexical diversity (pairwise intersections of lexicons are 13.7% on average), in the ratio of different parts of speech, in the quality of sentiment classification on their basis (the difference between the best and the worst is 0.074).

The created union lexicon (containing 20,401 words), which most fully reflects the sentiment lexica of the Russian language, is of considerable interest to researchers.

Further study of the issue would deal with, firstly, the creation of a compact effective sentiment lexicon based on the union lexicon and feature selection methods, secondly, the study of the quality of sentiment lexicons using lexicon-based sentiment classification techniques that can be used without training procedure.

---

[6] The union lexicon is not shown in Fig. 1.

Acknowledgments. This work was carried out as a part of the project of Government Order No. 34.2092.2017/4.6 of the Ministry of Education and Science of the Russian Federation.

References

1. Baccianella, S., Esuli, A., Sebastiani, F.: SentiWordNet 3.0: an enhanced lexical resource for sentiment analysis and opinion mining. In: Proceedings of the Seventh Conference on International Language Resources and Evaluation (LREC 2010), Valletta, pp. 2200–2204 (2010)
2. Balahur, A., Hermida, J.M., Montoyo, A.: Detecting implicit expressions of emotion in text: a comparative analysis. Decis. Support Syst. **53**, 742–753 (2012)
3. Blinov, P.D., Klekovkina, M.V., Kotelnikov, E.V., Pestov, O.A.: Research of lexical approach and machine learning methods for sentiment analysis. In: Computational Linguistics and Intellectual Technologies: Papers from the Annual International Conference "Dialogue-2013", vol. 12, no. 19, pp. 51–61 (2013)
4. Chen, Y., Skiena, S.: Building sentiment lexicons for all major languages. In: Proceedings of the 52nd Annual Meeting of the Association for Computational Linguistics, Baltimore, pp. 383–389 (2014)
5. Chetviorkin I., Loukachevitch N.: Extraction of Russian sentiment lexicon for product meta-domain. In: Proceedings of COLING 2012, Mumbai, pp. 593–610 (2012)
6. Habernal, I., Ptáček, T., Steinberger, J.: Supervised sentiment analysis in Czech social media. Inf. Process. Manag. **51**(4), 532–546 (2015)
7. Hailong, Z., Wenyan, G., Bo, J.: Machine learning and lexicon based methods for sentiment classification: a survey. In: Proceedings of the 11th Web Information System and Application Conference, Tianjin, pp. 262–265 (2014)
8. Hemmatian, F., Sohrabi, M.K.: A survey on classification techniques for opinion mining and sentiment analysis. Artif. Intell. Rev. 1–51 (2017)
9. Kiselev, Y., Braslavski, P., Menshikov, I., Mukhin, M., Krizhanovskaya, N.: Russian Lexicographic landscape: a tale of 12 dictionaries. In: Computational Linguistics and Intellectual Technologies: Papers from the Annual International Conference "Dialogue-2015", vol. 14, no. 21, pp. 254–271 (2015)
10. Koltsova, O.Yu., Alexeeva, S.V., Kolcov, S.N.: An opinion word lexicon and a training dataset for russian sentiment analysis of social media. In: Computational Linguistics and Intellectual Technologies: Papers from the Annual International Conference "Dialogue-2016". vol. 15, no. 22, pp. 277–287 (2016)
11. Kotelnikov, E., Bushmeleva, N., Razova, E., Peskisheva, T., Pletneva, M.: Manually created sentiment lexicons: research and development. In: Computational Linguistics and Intellectual Technologies: Papers from the Annual International Conference "Dialogue-2016", vol. 15, no. 22, pp. 300–314 (2016)
12. Linguistic Inquiry and Word Counts. http://liwc.wpengine.com. Accessed 20 May 2018
13. Liu, B.: Opinion Mining, Sentiment Analysis, and Opinion Spam Detection. https://www.cs. uic.edu/∼liub/FBS/sentiment-analysis.html. Accessed 20 May 2018
14. Loukachevitch, N., Dobrov, B.: RuThes linguistic ontology vs. russian wordnets. In: Proceedings of the 7th Global Wordnet Conference (GWC 2014), Tartu, pp. 154–162 (2014)
15. Loukachevitch, N., Levchik, A.: Creating a general russian sentiment lexicon. In: Proceedings of Language Resources and Evaluation Conference LREC-2016, pp. 1171–1176 (2016)

16. Mohammad, S.M., Turney, D.P.: Crowdsourcing a word-emotion association lexicon. Comput. Intell. **29**(3), 436–465 (2013)
17. Ohana, B., Tierney, B., Delany, S.-J.: Domain independent sentiment classification with many lexicons. In: 2011 IEEE Workshops of International Conference on Advanced Information Networking and Applications (WAINA), Singapore, pp. 632–637 (2011)
18. Pedregosa, et al.: Scikit-learn: machine learning in python. JMLR **12**, 2825–2830 (2011)
19. Potts, Ch.: Sentiment symposium tutorial: lexicons. In: Sentiment Analysis Symposium, San Francisco, 8–9 November 2011 (2011)
20. Qiu, G., Liu, B., Bu, J., Chen, C.: Opinion word expansion and target extraction through double propagation. Comput. Linguist. **37**(1), 9–27 (2011)
21. Saif, H., He, Y., Fernandez, M., Alani, H.: Contextual semantics for sentiment analysis of Twitter. Inf. Process. Manag. **52**(1), 5–19 (2016)
22. Stone, P.J., Dunphy, D.C., Smith, M.S., Ogilvie, D.M.: The General Inquirer: A Computer Approach to Content Analysis. MIT Press Cambridge, Cambridge (1966)
23. Taboada, M.: Sentiment analysis: an overview from linguistics. Annu. Rev. Linguist. **2**, 325–347 (2016)
24. Taboada, M., Brooke, J., Tofiloski, M., Voll, K., Stede, M.: Lexicon-based methods for sentiment analysis. Comput. Linguist. **37**(2), 267–307 (2011)
25. Tutubalina, E.V.: Metody izvlecheniya i rezyumirovaniya kriticheskih otzyvov pol'zovatelej o produkcii (Extraction and summarization methods for critical user reviews of a product). Kazan Federal University, Kazan (2016)
26. Wilson, T., Wiebe, J., Hoffmann, P.: Recognizing contextual polarity in phrase-level sentiment analysis. In: Proceedings of the 2005 Human Language Technology Conference and the Conference on Empirical Methods in Natural Language Processing (HLT/EMNLP-05), Vancouver, pp. 347–354 (2005)

Acoustic Features of Speech of Typically Developing Children Aged 5–16 Years

Alexey Grigorev, Olga Frolova, and Elena Lyakso[✉]

Saint Petersburg State University, Saint Petersburg, Russia
lyakso@gmail.com

Abstract. The study is aimed at investigating the formation of acoustic features of speech in typically developing (TD) Russian-speaking children. The purpose of the study is to describe the dynamics of the temporal and spectral characteristics of the words of 5–16 years old children depending on their gender and age. The decrease of stressed and unstressed vowels duration from child's words to the age of 13 years is revealed. Pitch values of vowels from words significantly decrease to the age of 14 years in girls and to the age of 16 years in boys. Pitch values of vowels from girls' words are higher vs. corresponding features from boys' words. Differences in the pitch values and vowel articulation index in boys and girls in different ages are shown. The obtained data on the acoustic features of the speech of TD children can be used as a normative basis in artificial intelligence systems for teaching children, for creating alternative communication systems for children with atypical development, for automatic recognition of child speech.

Keywords: Child speech · Acoustic features of speech · Pitch
Vowel articulation index

1 Introduction

The data on the acoustic features of infant's vocalizations [1, 2], speech of children during the first years of life [3, 4] are obtained for different languages. Acoustic characteristics of speech of preschoolers [5–7], junior schoolchildren, and teenagers [8–12] are less studied. The focus of research is shifted to investigating child's speech disorders [13–15].

The works of Child speech research group of Saint Petersburg State University describe the acoustic features of vocalizations and speech of Russian typically developing (TD) infants [16, 17] compared with orphans and children with neurological disorders [18, 19], the dynamics of the acoustic features of vowels in vocalizations and words of TD children [20] and twins [21] from birth to the age of 7 years. The comparative study of speech formation in TD children and children with autism spectrum disorders (ASD) is carried out [22, 23]. However, in these studies, the speech of TD children was recorded in the model situations - the repetition of words [22, 23] and under the different emotional states [24]. Information on the acoustic characteristics of calm (neutral) speech of Russian-speaking TD children in the age range of 5–16

© Springer Nature Switzerland AG 2018
D. Ustalov et al. (Eds.): AINL 2018, CCIS 930, pp. 152–163, 2018.
https://doi.org/10.1007/978-3-030-01204-5_15

years is absent. Data on the acoustic characteristics of child's speech is necessary in clinical practice to clarify the diagnosis, for automatic speech recognition systems [25].

The aim of the study is to determine the age dynamics of temporal and spectral characteristics of vowels in words in 5–16 years old Russian children, depending on the gender and the age of the child.

2 Method

2.1 Data Collection

240 children aged 5–16 years (10 boys and 10 girls in each age) participated in the study. According pediatricians' conclusion all children developed normally, did not have diagnosed hearing and speech disorders. All children were born and have been living in St. Petersburg city with parents who were also born in St. Petersburg or have been living there for more than 10 years. For all the children the first language (L1) was Russian. At school, children were taught the second language (L2) English.

Audio records of child speech with parallel records of child's behavior in the situation of dialogue with the experimenter were made. The standard set of experimenter's questions addressed to the child was used. The experimenter began the dialogue with the request to say your name and age. Then the experimenter consistently asked questions:

- Do you like to go to school/kindergarten?
- What do you like in school/kindergarten (classes or play with friends)?
- What are your favorite tasks? Why?
- Do you have any hobbies?
- What are your favorite movies, cartoons, books, games (computer/desktop/mobile)?
- Do you have brothers or sisters?
- Do you have pets?
- Did you visit the zoo, circus, and museum?

This set of questions allowed obtaining the child's replicas containing similar and identical words. For example every child used in replicas the words: /like (nrAvitsya) – do not like (ne nrAvitsya)/, /know (znAyu) – do not know (ne znAyu)/, /Russian (rUsskiy)/, /bored (skUchno)/, /plays (Igry)/, /tiger (tIgr)/.

The duration of the dialogues was 5–10 min. The recordings were made by the "Marantz PMD660" recorder with a "SENNHEIZER e835S" external microphone and camera "SONY HDR-CX560E". Speech files are stored in Windows PCM format, 44100 Hz, 16 bits per sample.

All procedures were approved by the Health and Human Research Ethics Committee and signed informed consent was obtained from parents of the child participant.

2.2 Data Analysis

Spectrographic analysis of speech was carried out in the "Cool Edit Pro" sound editor (Syntril. Soft. Corp. USA). We analyzed the duration of stressed and unstressed vowels and the stationary part of vowels; pitch values, formants frequencies (F1, F2) for the stationary part of vowels.

The acoustic features reflecting the basic physiological processes in the vocal tract during the articulation of the speech signal were chosen. Temporal features (vowels duration) are associated with the formation of speech breathing, the pitch values are the indicator of the frequency of oscillations of the vocal folds, the values of the two first formants reflect the processes occurring in the oral cavity and are acoustic keys for the identification of vowels.

Formant triangles with apexes corresponding to the vowels /a/, /u/, and /i/ in F1, F2 coordinates were plotted and their areas were compared. Vowels formant triangle areas [20] and vowel articulation index (VAI) [26] were calculated.

3 Result and Discussion

3.1 Duration of Vowels from the Words of 5–16 Years Old Children

We found dynamics of the duration of stressed and unstressed vowels in child's words. Duration of stressed vowels in girls' words significantly increases from the age of 5 years to 7 years (Mann–Whitney test), decreases during the age of 9–11 years ($p < 0.01$ Kruskal–Wallis test), and stabilizes at the age of 13–16 years ($p < 0.001$) (Fig. 1A).

The stressed vowels duration in the boys' words is significantly higher than the corresponding values of unstressed vowels duration for all ages except the age of 7 years. ($p < 0.001$ for child's age of 5 years, 8–9 years, 11 years and from 13 to 16 years; $p < 0.01$ for age of 6 years, 10 years and 12 years, Mann–Whitney test). The duration of stressed vowels in the words of boys reduces to the age of 13 years ($p < 0.05$) (Fig. 1B). The duration of unstressed vowels in the words of the girls reduces to the age of 13–16 years ($p < 0.05$) and to the age of 13 years in the words of the boys ($p < 0.05$).

Girls age correlates with the duration of stressed vowels $F(1.798) = 83.608$; $p < 0.000$ (Beta = -0.308; $R^2 = 0.095$) and the stationary part of stressed vowels $F(1.798) = 315.61$; $p < 0.000$ (Beta = -0.532; $R^2 = 0.283$); with the duration of unstressed vowels $F(1.1361) = 63.295$; $p < 0.000$ (Beta = -0.211; $R^2 = 0.044$) and the stationary part of unstressed vowels $F(1.1361) = 488.50$; $p < 0.000$ (Beta = -0.514; $R^2 = 0.264$) – Regression analysis.

Boys age correlates $F(1.760) = 119.26$; $p < 0.000$ with the duration of stressed vowels (Beta = -0.368; $R^2 = 0.136$) and the stationary part of stressed vowels $F(1.760) = 435.45$; $p < 0.000$ (Beta = -0.604; $R^2 = 0.364$).

The data on the stabilization of stressed and unstressed vowels duration to the age of 13 years may indicate the formation of speech breathing to this age.

According to the literature, speech breathing in children at the age of 7 years differs from speech breathing of adults [8]. These differences pass away by the age of 10 years, but developing some features of speech breathing (respiratory volume of lungs, sound pressure level) continues in adolescence [8]. Speech breathing features depend on the age of the informant, but not the informant's gender [27].

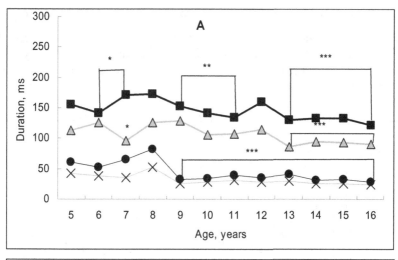

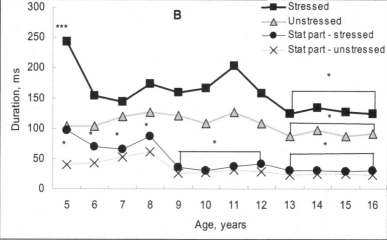

Fig. 1. The duration of stressed and unstressed vowels from the words of girls (A) and boys (B) aged 5–16 years. Square marker – stressed vowels, triangle marker - unstressed vowels, round marker – stationary part of stressed vowels, X-marker – stationary part of unstressed vowels. * - p < 0.05; ** - p < 0.01; *** - p < 0.001, Kruskal–Wallis test. Horizontal axis – age, years; vertical axis – duration, ms.

Our data on the stabilization of stressed and unstressed vowels duration up to the age of 13 years may point at the end of speech breathing developing at this age.

Thus, we revealed the main trends of the vowels duration. The identification of specific correlation between sex, age, and nonlinear variation in the vowels duration will be the subject of further work.

3.2   Pitch Values of Vowels from the Words of 5–16 Years Old Children

Stressed and unstressed vowels pitch values dynamics with child's age is traced. Pitch values of stressed vowels in girls' words reach maximum at the age of 5 years and decrease till 9 years and remain stable at the age of 9–13 years, decrease to the age of 14–16 years (Fig. 2A, Tables 1, 2).

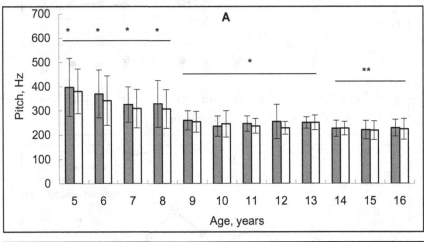

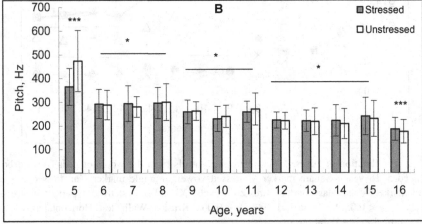

Fig. 2. Pitch values of vowels from 5–16 years olds' words. A – girls, B – boys, * - $p < 0.05$; ** - $p < 0.01$; *** - $p < 0.001$ – Mann–Whitney test. Horizontal axis – age, years; vertical axis – pitch values, Hz.

Figure 2B shows the same data for boys. Maximal values of pitch of stressed vowels are revealed in the words of 5 years old boys, after pitch decreases to the age of 6–8 years, 9–11 years, 12 years. Minimal values of pitch are revealed at the boy's age of 16 years. Tables 1, 2 represents accurate age and gender data of pitch values of child's vowels in words.

Table 1. Pitch values of stressed vowels, Hz.

Age, years	Boys	Girls
5	365 ± 78	398 ± 119
6	292 ± 61	370 ± 99
7	294 ± 75	326 ± 73
8	296 ± 66	329 ± 96
9	259 ± 50	261 ± 39
10	229 ± 53	237 ± 42
11	259 ± 44	247 ± 31
12	224 ± 34	256 ± 71
13	221 ± 45	252 ± 24
14	222 ± 66	227 ± 33
15	240 ± 79	222 ± 38
16	186 ± 48	230 ± 35

Table 2. Pitch values of unstressed vowels, Hz.

Age, years	Boys	Girls
5	474 ± 129	381 ± 92
6	288 ± 62	343 ± 102
7	280 ± 43	310 ± 79
8	301 ± 77	308 ± 80
9	263 ± 39	256 ± 43
10	240 ± 47	247 ± 54
11	271 ± 67	238 ± 30
12	221 ± 35	230 ± 26
13	218 ± 56	252 ± 30
14	208 ± 64	228 ± 28
15	230 ± 75	219 ± 38
16	175 ± 50	225 ± 43

Pitch values of stressed vowels from girls' words are significantly higher vs. corresponding values of vowels from boys' words at the age of 6 years ($p < 0.001$; Mann–Whitney test), 7–8 years ($p < 0.05$), 13 years ($p < 0.01$), and 16 years ($p < 0.001$). Pitch values of unstressed vowels from girls' words are significantly higher vs. corresponding values of vowels from boys' words at the age of 6 years ($p < 0.01$), 10 years ($p < 0.05$), 13–14 years ($p < 0.001$), and 16 years ($p < 0.001$). Pitch values of unstressed vowels from boys' words are significantly higher than corresponding values of vowels from girls' words at the age of 5 years and 11 years ($p < 0.05$).

Child's gender correlates $F(6.1797) = 5.155$; $p < 0.0000$ with pitch values of stressed vowels (Wilks'Lambda 0.965; $p = 0.0000$); $F(6.2669)$; $p < 0.0000$ with pitch values of unstressed vowels (Wilks'Lambda 0.974; $p < 0.0000$) – Discriminant analysis.

Girls' age correlates with pitch values of stressed vowels $F(1.892) = 458.99$; $p < 0.000$ (Beta = -0.583; $R^2 = 0.34$) and pitch values of unstressed vowels $F(1.1361) = 643.32$; $p < 0.000$ (Beta = -0.566; $R^2 = 0.321$) – Regression analysis. Boys' age correlates with pitch values of stressed vowels $F(1.808) = 174.49$; $p < 0.000$ (Beta = -0.421; $R^2 = 0.178$).

The results on the decrease of pitch values with child's age correspond to the data for other languages [10, 12] and reflect the general patterns of voice formation in ontogenesis [28–32]. In our study sharp changes in the pitch values in boys at the ages of 6 years, 9 years, 12 years and 16 years were revealed. More linear decrease of pitch values with child's age was described in girls vs. boys. Age-related anatomical changes in the vocal tract, in particular, changes in its length could be used as an explanation of these data. Differences in the length of the vocal tract between boys and girls after the age of 12 years revealed by MRI data caused changes in the pitch values between girls and boys [29]. The authors concluded based on the fMRI that boys and girls have different age dynamics in the vocal tract length, causing various dynamics of pitch [32]. Two age periods with sharp decreases in the pitch values 6–8 years and 12–15 years are revealed in boys, all pitch changes in girls are more linear without sudden changes [32].

3.3 Formant Characteristics of Vowels from the Words of 5–16 Years Old Children

Figure 3 presents the formant triangles with apexes corresponding to values of the first and second formants of stressed vowels /a/, /i/, /u/ from girls (A) and boys (B) words. The first formant values of stressed vowel /a/ from girls' words are maximal at the age of 5 years and decrease to the age of 12 years ($p < 0.05$; Kruskal–Wallis test). The first formant values of stressed vowels /a/, /u/, /i/ from boys' words are maximal at the age of 5 years and decrease with child's age. The second formant values of stressed vowels from girls' and boys' words show tendency to decrease with child age.

The formant triangles of stressed vowels shift to the low-frequency region on the two-formant coordinate plane by the values of the first formant – up to the age of 14 years for girls and up to the age of 16 years for boys (Fig. 3).

VAI of stressed vowels from girls' and boys' words changes with children's age non-linearly. The end of the preschool period is characterized by the maximum values of VAI (for boys at the age of 6 years, for girls at the age of 7 years), which can be explained by preparing children for schooling and the need for active use of verbal communication in the learning process. At the age of 8 years VAI values are high and similar for boys and girls. A further decrease in VAI may be due to the increase in fluency of speech and the termination of the articulation skills mastering.

VAI of stressed vowels from girls' words is 0.75 (conventional units) at the age of 5 years and 0.82 at the age of 16 years. VAI of stressed vowels from boy's words is 0.61 at the age of 5 years and 0.87 at the age of 16 years (Fig. 4).

The values of formant triangle areas of stressed vowels from girls' words are higher vs. boys' words except the ages of 10, 11, 12 years (Fig. 5).

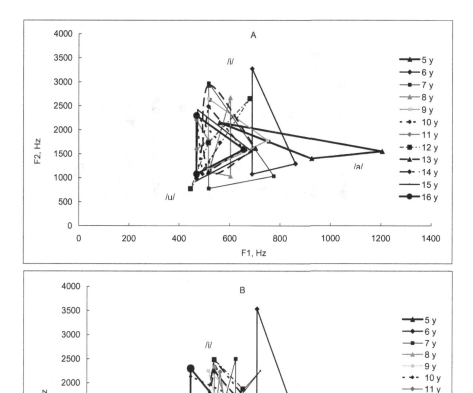

Fig. 3. The stressed vowels formant triangles with apexes /a/, /u/, /i/ from words of girls (A) and boys (B). Horizontal axis values are F1, Hz, vertical axis values are F2, Hz. Bold lines indicate the data for 5 and 16 years old children (boundaries of the analyzed age range).

Revealed differences in the values of VAI and formant triangle areas with the age of boys and girls are indirectly confirmed by the data on the material of the Chinese language about the significant decrease in the values of the first two formants in the speech of children in the age range 3–18 years [33]. The authors associate differences between boys and girls on the base of the vowels' formant frequencies with differences in the volume of the pharynx of children [33].

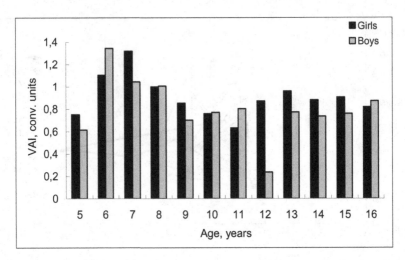

Fig. 4. Articulation index of stressed vowels. Black columns – data for girls, grey columns – data for boys. Horizontal axis – age of child, years, vertical axis - vowel articulation index values, conventional units.

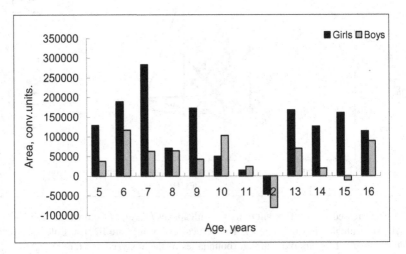

Fig. 5. Areas of vowels formant triangles. Black columns – data for girls, grey columns – data for boys. Horizontal axis – age of child, years, vertical axis - areas of vowels formant triangles values, conventional units.

3.4   Phonetic Data

The phonetic analysis revealed that children use normative phonemes for Russian language in words at the age of 8–16 years. Some differences in pronunciation related to individual features of the child and they do not affect the lexical meaning of the word. Unformed phonemes /l/, /ʃ/, /ʃ'/, /tS'/, /r/, /Z/, /s/, /s'/ were described in 5–7 years old children. Phonetic analysis revealed sound replacements not affecting the lexical meaning of the word: changes and replacements of phonemes /r'/ and /r/, change /ʃ/ to /s/, and /tS'/ to /t'/, replacements of group of consonants with one phoneme.

4 Conclusions

Age dynamics of the duration, pitch values, and spectral features of stressed and unstressed vowels from words of Russian-speaking boys and girls aged 5–16 years was described. Age dynamic of the duration of stressed and unstressed vowels and their stationary parts was shown. Duration of stressed and unstressed vowels from child's words significantly decrease to the age of 13 years in children of both genders. Significant decrease of pitch values of vowels from words with child's age was shown. The difference in pitch values between girls and boys was defined. These data reflect general patterns of the voice formation in ontogenesis, taking into account the gender of the informant. The clarity of articulation of Russian-speaking 5–16 years old children was described by the vowels articulation index.

The obtained data on the acoustic features of the speech of TD children can be used as a normative basis in artificial intelligence systems for teaching children, for creating alternative communication systems for children with atypical development, for automatic recognition of child speech [34], for teaching children the clarity of pronunciation based on the use of acoustic feedback.

Acknowledgements. This study is financially supported by Russian Science Foundation (Project № 18-18-00063).

References

1. Kuhl, P.K., Meltzoff, A.N.: Infant vocalizations in response to speech: vocal imitation and developmental change. J. Acoust. Soc. Am. **100**(4), 2425–2438 (1996)
2. Nathani, S., Ertmer, D.J., Stark, R.E.: Assessing vocal development in infants and toddlers. Clin. Linguist. Phon. **20**(5), 351–369 (2006)
3. Lyakso, E., Gromova, A., Frolova, O., Romanova, O.: Acoustic aspect of the formation of speech in children in the third year of life. Neurosci. Behav. Physiol. **35**(6), 573–583 (2005)
4. Liu, S., Jin, Y., Yu, H., Yang, L.: Study on the acoustic characteristics of speech and physiological development of vocal organs for two-year-old children. In: Fifth International Conference on Instrumentation and Measurement, Computer, Communication and Control, pp. 576–579. IEEE, Qinhuangdao (2015)
5. Ballard, K.J., Djaja, D., Arciuli, J., James, D.J.H., van Doorn, J.: Developmental trajectory for production of prosody: lexical stress contrastivity in children 3 to 7 years and adults. J. Speech Lang. Hear. Res. **55**, 1716–1735 (2012)
6. Brehm, S.B., Weinrich, B.D., Sprouse, D.S., May, S.K., Hughes, M.R.: An examination of elicitation method on fundamental frequency and repeatability of average airflow measures in children age 4:0–5:11 years. J. Voice **26**(6), 721–725 (2012)

7. Lyakso, E.E., Frolova, O.V., Grigorev, A.S.: Infant vocalizations at the first year of life predict speech development at 2–7 years: longitudinal study. Child Psychol. **5**(12), 1433–1445 (2014)

8. Hoit, J., Hixon, T., Watson, P., Morgan, W.: Speech breathing in children and adolescents. J. Speech Hear. Res. **33**, 51–69 (1990)

9. Vorperian, H., Kent, R.: Vowel acoustic space development in children: a synthesis of acoustic and anatomic data. J. Speech Lang. Hear. Res. **50**(6), 1510–1545 (2007)

10. Lee, S., Iverson, G.K.: The development of monophthongal vowels in Korean: age and sex differences. Clin. Linguist. Phon. **22**(7), 523–536 (2008)

11. Harries, M., Hawkins, S., Hacking, J., Hughes, I.: Changes in the male voice at puberty: vocal fold length and its relationship to the fundamental frequency of the voice. J. Laryngol. Otol. **112**(5), 451–454 (1998)

12. Lee, S., Potamianos, A., Narayanan, S.: Acoustics of children's speech: developmental changes of temporal and spectral parameters. J. Acoust. Soc. Am. **105**, 1455–1468 (1999)

13. Bonneh, Y.S., Levanon, Y., Dean-Pardo, O., Lossos, L., Adini, Y.: Abnormal speech spectrum and increased pitch variability in young autistic children. Front. Hum. Neurosci. **4** (237), 1–7 (2011)

14. Kent, R.D., Vorperian, H.K.: Speech impairment in down syndrome: a review. J. Speech Lang. Hear. Res. **56**(1), 178–210 (2013)

15. Fortunato-Tavares, T., Andrade, C.R.F., Befi-Lopes, D., Limongi, S.O., Fernandes, F.D.M., Schwartz, R.G.: Syntactic comprehension and working memory in children with specific language impairment, autism or down syndrome. Clin. Linguist. Phon. **29**(7), 499–522 (2015)

16. Lyakso, E.: Characteristics of infant's vocalizations during the first year of life. Int. J. Psychophysiol. **30**, 150–151 (1998)

17. Lyakso, E.E.: Phonological developments of Russian children during the first postnatal year. In: Joint Conference of the IX International Congress for the Study of Child Language and the Symposium on Research in Child Language Disorders, p. 171. SRCLD-IASCL, Madison, USA (2002)

18. Lyakso, E., Frolova, O.: Russian vowels system acoustic features development in ontogenesis. In: Interspeech 2007, pp. 2309–2313. Antwerp, Belgium (2007)

19. Lyakso, E.E., Frolova, O.V., Stolyarova, E.I.: Russian child speech development in mother–child interaction: basis rules and individual features. Int. J. Psychophysiol. **69**(3), 306–307 (2008)

20. Lyakso, E.E., Grigor'ev, A.S.: Dynamics of the duration and frequency characteristics of vowels during the first seven years of life in children. Neurosci. Behav. Physiol. **45**(5), 558–567 (2015)

21. Kurazova, A.V., Lyakso, E.E.: Speech development of children and vocal-speech interaction in the triads "mother-twins": longitudinal study. Bull. St. Petersburg State Univ. **3**, 93–103 (2012). (in Russian)

22. Lyakso, E., Frolova, O., Grigorev, A.: A comparison of acoustic features of speech of typically developing children and children with autism spectrum disorders. Lect. Notes Comput. Sci. **9811**, 43–50 (2016)

23. Lyakso, E., Frolova, O., Grigorev, A.: Perception and acoustic features of speech of children with autism spectrum disorders. Lect. Notes Comput. Sci. **10458**, 602–612 (2017)

24. Lyakso, E.E., Frolova, O.V., Grigor'ev, A.S., Sokolova, V.D., Yarotskaya, K.A.: Recognition by adults of emotional state in typically developing children and children with autism spectrum disorders. Neurosci. Behav. Physiol. **47**(9), 1051–1059 (2017)

25. Fernando, S., et al.: Automatic recognition of child speech for robotic applications in noisy environments. Comput. Speech Lang. 1–28 (2016). arXiv:1611.02695v1 [cs.CL]

26. Roy, N., Nissen, S.L., Dromey, C., Sapir, S.: Articulatory changes in muscle tension dysphonia: evidence of vowel space expansion following manual circumlaryngeal therapy. J. Commun. Disord. **42**(2), 124–135 (2009)
27. Boliek, C.A., Hixon, T.J., Watson, P.J., Jones, P.B.: Refinement of speech breathing in healthy 4- to 6-year-old children. J. Speech Lang. Hear. Res. **52**, 990–1007 (2009)
28. Hacki, T., Heitmuller, S.: Development of the child's voice: premutation, mutation. Int. J. Pediatr. Otorhinolaryngol. **49**(1), 141–144 (1999)
29. Fitch, W.T., Giedd, J.: Morphology and development of the human vocal tract: a study using magnetic resonance imaging. J. Acoust. Soc. Am. **106**, 1511–1522 (1999)
30. Hollien, H., Green, R., Massey, K.: Longitudinal research on adolescent voice change in males. J. Acoust. Soc. Am. **96**, 2646–2654 (1994)
31. Perry, T.L., Ohde, R.N., Ashmead, D.H.: The acoustic bases for gender identification from children's voices. J. Acoust. Soc. Am. **109**, 2988–2998 (2001)
32. Markova, D., et al.: Age- and sex-related variations in vocal-tract morphology and voice acoustic during adolescence. Horm. Behav. **81**, 84–96 (2016)
33. Wan, P., Huang, Z., Zheng, Q.: Acoustic elementary research on voice resonance of Chinese population. Chin. J. Surg. **24**(6), 250–252 (2010)
34. Kaya, H., Salah, A.A., Karpov, A., Frolova, O., Grigorev, A., Lyakso, E.: Emotion, age, and gender classification in children's speech by humans and machines. Comput. Speech Lang. **46**, 268–283 (2017)

Social Interaction Analysis

Profiling the Age of Russian Bloggers

Tatiana Litvinova[1,2(✉)] ⓘ, Alexandr Sboev[2] ⓘ,
and Polina Panicheva[1] ⓘ

[1] Voronezh State Pedagogical University, 86 Lenina ul.,
Voronezh 394043, Russia
`centr_rus_yaz@mail.ru`
[2] National Research Center "Kurchatov Institute", 1, Akademika Kurchatova pl.,
Moscow 123182, Russia

Abstract. The task of predicting demographics of social media users, bloggers and authors of other types of online texts is crucial for marketing, security, etc. However, most of the papers in authorship profiling deal with author gender prediction. In addition, most of the studies are performed in English-language corpora and very little work in the area in the Russian language. Filling this gap will elaborate on the multi-lingual insights into age-specific linguistic features and will provide a crucial step towards online security management in social networks. We present the first age-annotated dataset in Russian. The dataset contains blogs of 1260 authors from *LiveJournal* and is balanced against both age group and gender of the author. We perform age classification experiments (for age groups 20–30, 30–40, 40–50) with the presented data using basic linguistic features (lemmas, part-of-speech unigrams and bigrams etc.) and obtain a considerable baseline in age classification for Russian. We also consider age as a continuous variable and build regression models to predict age. Finally, we analyze significant features and provide interpretation where possible.

Keywords: Authorship profiling · Age prediction · Russian language
Text classification

1 Introduction

Determining demographic characteristics (gender, age) of the authors of online texts is crucial in many areas such as business intelligence and digital forensics. Gender profiling is one of the most developed areas in this field. However, recent works in this area deal with age detection as well. Age identification of online authors is in high practical demand in online security applications. On the other hand, it contributes to the overall understanding of human idiolect features. Work on age author profiling has been performed in a variety of languages. However, to our knowledge there has been very little work in the area in the Russian language, despite a number of successful approaches to profiling of gender (Litvinova et al. 2017, 2018; Panicheva et al. 2018; Sboev et al. 2016, 2018) as well as different psychological characteristics (Litvinova et al. 2016).

This paper is aimed at prediction of the age of the authors of Russian-language blogs using different combinations of linguistic features and two different approaches.

D. Ustalov et al. (Eds.): AINL 2018, CCIS 930, pp. 167–177, 2018.
https://doi.org/10.1007/978-3-030-01204-5_16

We consider age prediction task as a text classification task as well as a regression problem. To the best of our knowledge, this paper is the first to address the problem of predicting age of the authors of texts in Russian.

The paper is organized as follows. Section 2 surveys the literature on age prediction. In Sect. 3 we present our data, features, machine learning algorithms, and evaluation setting. In Sect. 4 we present our classification results and perform feature analysis. We conclude this paper in Sect. 5 by summarizing our study.

2 State of the Art

One of the first works aimed at predicting the age of blog authors based on linguistic parameters is that by Schler et al. (2006), which examines bloggers based on their age at the time of the experiment, whether in the 10's, 20's or 30's age bracket. They have created "Blog Authorship Corpus" which is publicly available and is used for many more recent studies and PAN competition (see below). They have identified interesting changes in content and style features across categories, in which they include blogging words (e.g., "LOL"), all defined by the Linguistic Inquiry and Word Count (LIWC) software (Pennebaker et al. 2001). Their approach allows to distinguish between bloggers in the 10's and in the 30's with a relatively high accuracy (above 96%), distinguishing 10 s from 20 s is also achievable with accuracy of 87.3%, but many 30 s are misclassified as 20 s, which results in an overall accuracy of 76.2%. Their work shows that ease of classification is dependent in part on the age group distinction.

Argamon et al. (2007) used the same corpus, in order to refine in the gender and age identification task (using three classes: 10 s, 20 s, or 30+). This study is the first one to link together earlier observations regarding age-linked and gender-linked writing variation that have not previously been connected. The authors found out that the same features are useful for predicting age and gender. Using function words they achieved best accuracy of 69.4%, while using just the high information–gain words the authors obtained best accuracy of 76.2%. They concluded that topic preference is most related to blogger age, although there is definitely a marked effect on writing style as well.

Blogs were in the spotlight in a paper (Rosenthal and McKeown 2011). The authors approach age prediction by identifying a shift in writing style over a 14 year time span from birth years 1975–1988, so they perform binary classification between blogs before and in/after each year in this range. They motivated examining precisely these years due to the emergence of social media technologies during that time. The best results reached an accuracy of 79.96% and 81.57% for 1979 and 1984 respectively using bag-of-words, personal interests, online behavior, and lexical-stylistic features.

Age detection has typically been modeled as a classification problem, although this approach often suffers from ad hoc and dataset dependent age boundaries (Rosenthal and McKeown 2011). In contrast, recent works have also explored predicting age as a continuous variable. For example, the authors in Nguyen et al. (2011) used linear regression to predict author age. They used a three datasets with different characteristics (including above mentioned Blog corpus by Schler et al. 2006). They used word unigrams and part-of-speech (POS) unigrams and bigrams as features. Additionally they used the LIWC tool to extract features. The authors obtained correlations up to

0.74 and mean absolute errors between 4.1 and 6.8 years. They concluded that even a unigram only baseline already gives strong performance and many POS patterns are strong indicators of old age.

Most of the previous studies use word unigrams as baseline features. Additional processing was used like POS tagging and LIWC tool to categorize words.

Since 2013, a lot of relevant research has been published in the context of the shared task on author profiling organized at PAN.[1] The participants used several different feature types to approach the problem of age and gender profiling in blogs, social media and other sources: content-based (bag of words, word n-grams, term vectors, named entities, dictionary words, slang words, contractions, sentiment words, etc.) and stylistic-based features (punctuations, POS, Twitter-specific elements, readability measures, and so forth), however very few studies perform detailed feature analysis. In the 2016 PAN shared task, age was divided into 5 groups, (a) 18–24; (b) 25–34; (c) 35–49; (d) 50–64; (e) 65+, with distances between real and predicted classes 0.6951 for English blogs (SD 0.7199) and 0.8176 for Spanish blogs (SD = 0.8775) (Rangel et al. 2016).

Most of the work related to age prediction focuses on predicting the age group of the author in terms of "young" versus "adult". To sum up, it was found that younger people use more capitalization of words, shorter words and sentences, more self-references, more slang words, and more Internet acronyms (see Nguyen et. al. 2016, for review), while older people use more first-person plural pronouns (we), prepositions, determiners, articles, longer words, longer sentences, links, hash tags. However, as was shown in Nguyen et al. (2013, 2017), strong changes in language style take place in the younger ages; after an age of around 30 most variables show little change, and therefore it is harder to predict the correct age of older people. To improve the accuracy of an age prediction model, one should pay closer attention to identifying variables that show more change at older ages. Not only predictive models, but also humans identify older people poorly (Nguyen et al. 2014), with prediction errors already starting for the late 20 s, and the gap between actual and predicted age increasing with age. One plausible explanation for this fact is that people between 30 and 55 years use more standard forms as they experience the maximum societal pressure to conform in the workplace (Nguyen 2017). Younger people and retired people, on the contrary, use more non-standard forms (Nguyen 2017).

Another interesting but understudied issue is an interrelation between age and other characteristics. For example, Argamon et al. (2007) have found out that the linguistic factors that increase in use with age (Articles, Prepositions, Religion, Politics, Business, and Internet) are just those used more by males of any age, and conversely, those that decrease in use with age are those used more by females of any age (Personal Pronouns, Conjunctions, Auxiliary Verbs, Conversation, At Home, Fun, Romance, and Swearing).

Most of the above mentioned research has been performed on texts in English. As of now, the Russian language remains understudied with respect to age profiling. We are aware of only several studies which deal with Russian-language texts with respect

---

[1] https://pan.webis.de/index.html (last accessed 2018/05/21).

to age prediction (Tutubalina and Nikolenko 2017; Alekseev and Nikolenko 2016; Gomzin et al. 2018). Two former studies make use of algorithms based on word embeddings. The authors of Tutubalina and Nikolenko (2017) deal with age group classification with 7 groups, based on medical texts. The authors of Alekseev and Nikolenko (2016) perform age regression, and have constructed their dataset from Odnoklassniki, a Russian social network. However, posts from this network contain very few personal linguistic content, - a fact which is out of scope of the study. They have obtained the results which are slightly (0.2 years) above a baseline, which is calculated as the mean age of a users friend (MAE 6.7). Both works employ elaborate neural network models with high performance in age prediction; however, most of the models are uninterpretable.

The latter work (Gomzin et al. 2018) is a recent attempt at classifying age group and education level of users of the social network Vkontakte based on their comments. They deal with 5 unbalanced age classes, and mostly address the task with linear models of word and character ngrams. Unfortunately, due to the larger number of classes, the unbalanced setting, and absence of linguistic interpretation, the results are not strictly comparable to our findings.

Neither of the works mentioned aims at providing linguistic insights into the language of people of different age.

3   Materials and Methods

3.1   Dataset

In our study, we use blogs from LiveJournal to build our research corpus. In the Russian Internet, LiveJournal, according to its statistics, has a significant audience, actually functioning as a popular social network. Many people known in Russia have their own "live journals", as well as parties, social movements and organizations.

We chose to use LiveJournal blogs for our corpus because the website provides an easy-to-use XML API. In addition, LiveJournal gives bloggers the opportunity to post their age in their profile.

First, we scraped a list of 120,000 most popular bloggers in Russian LiveJournal rating (https://www.livejournal.com/ratings/users/authority/?country=cyr) with the python Scrapy library (https://scrapy.org/). By using the LiveJournal 'friend of a friend' resource (http://exampleusername.livejournal.com/data/foaf.rdf) we collected their nickname, name, date of birth and a number of the latest posts in their rss feed (http://exampleusername.livejournal.com/data/rss). The rss feed only provides up to around 25 latest posts; however, it fits our task perfectly as we are interested in presenting a variety of different authors, with no single author dominating the corpus by a large number of posts.

We selected a list of users who provided their date of birth and first/last name which allowed to identify their age and gender. We used PyMorphy (Korobov 2015) to identify the gender of the first name; if this gave no result, we proceeded to a number of heuristics indicating a common male or female family name suffix, such as 'ova', 'eva', 'kaya', 'kaja', 'ina' for female names, 'ov', 'ev', 'kij', 'kiy', 'in' for male names, and their cyrillic counterparts. We filtered out authors if more than a half of their posts contained pictures, links, advertisement blocks or reposts. We only chose posts in the Russian language, as identified by the langid library (Lui and Baldwin 2012). We also deleted all the posts containing references to other posts or social networks, reposts containing links to the original text, posts shorter than 1,000 symbols, or if the calculated age of the author at the time of posting was above 80.

Thus we obtained 22,707 posts by 2,705 authors annotated with gender and age of the author at posting time. After filtering out all the posts older than 2014, the corpus contained 15,060 posts. We randomly chose 1,000 posts for age groups 20–30, 30–40, 40–50 from our initial dataset, obtaining a corpus of 6,000 posts (43 MB) by 1,196 unique authors balanced by age and gender, with some authors appearing in different groups, as their posts were written at different age intervals. All the following experiments are reported for the described balanced dataset of 6,000 posts (Fig. 1).

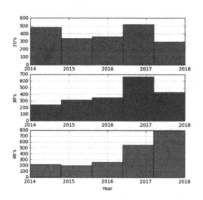
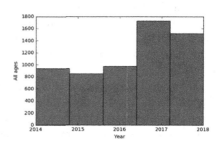

Fig. 1. Years' distribution of the posts

The mean age of bloggers was 35 years old (±7.6 y.o.). The number of texts by single author ranged from 1 to 24 with a mean of 5, std = 4.4. The mean size of texts ranged from 76 to 5,687 tokens, with a mean of 529 tokens (std = 577). The corpus is freely available on request.

3.2   Features

We have performed baseline age classification experiments with the blog corpus. Four groups of features were applied as well as their combinations (1991 in total):

- lemma unigrams and bigrams;
- POS unigrams and bigrams;
- QUITA features (Kubát et al. 2014) obtained on lemmatized corpus;
- grammatical tense, aspect and their combinations.

Lemmas, POS and tense features were identified by PyMorphy, with tokenization performed by happierfuntokenizer (http://www.wwbp.org/data.html) designed for social media data tokenization. QUITA features were identified with the Quantitative Text Analyzer tool. QUITA focuses mainly on indicators connected to the frequency structure of a text (type-token ratio, h-point, hapax legomenon percentage, and so on)

We only accounted for features and their bigrams which occurred in at least 5% of texts. L1-normalization was applied.

We also performed experiments with and without gender as a feature to evaluate its usefulness in age prediction.

3.3 Experiment Settings

We have performed two age prediction experiments:

1. Age prediction as text classification task with 3 groups: 20–30, 30–40, and 40–50 y.o.
2. Regression task, with age as a continuous variable in the range 20–50 y.o.

All the experiments were performed with 10-fold group cross-validation, i.e. texts by a single author always belonged to a single fold. It means that we partitioned our data to 10 even sized and random parts, and then used one part for validation and other 9 as training dataset. We did so 10 times and computed the overall evaluation for the entire dataset, which is equal to the weighted average among the cross-validation iterations.

The unit of classification is a blog post. In the age classification task we applied Support Vector Classification with L1 regularization (SVC) (one-vs-all), and Random Forest (RF) (multiclass classification). The baseline accuracy in the classification task is 0.33 – the frequency of the most common class, or any class in the balanced setting. We report accuracy and F-macro scores for the achieved results.

In the regression task we report the results of linear regression model with the same L1 (also known as Lasso) regularization, as the preliminary experiments with other regression methods performed near the 'dummy' baseline. The evaluation metrics are mean absolute error (MAE) and R-squared, or explained variance ratio (R2). The regression baseline MAE is estimated for the constant mean value age prediction (35 y. o.) and is 6.62 years.

L1 regularization was chosen as the feature selection method in both tasks, as (1) it only supports the features useful in the prediction task, eliminating highly inter-correlated features, which cannot be achieved by univariate feature selection methods; (2) L1 regularization allows for a very sparse resulting representation comparing to L2-regularization. In our high-dimensional task both effects are beneficial.

The experiments were performed with scikit-learn library (Pedregosa et al. 2011).

We have identified the significant features in the prediction tasks by the highest absolute value coefficients in the linear models and the mean feature importance in the Random Forest based on Gini impurity (http://scikit-learn.org). We also identified features significantly correlated with age by applying Student t-test in the classification and Pearson correlation in the regression tasks with Bonferroni-Holm multiple hypotheses correction (scipy library, Jones et al. 2014).

4   Results and Discussion

4.1   Prediction Results

The results for classification and regression tasks are presented below (Tables 1, 2 and 3). We only provide the best results obtained with the use of each classifier, since we have run many experiments with different settings (number of trees for RF, penalty and C for SVC). We also present the results of classification for male and female texts for the best set of parameters and full set of parameters.

We can see that SVC and RF performs similarly in terms of Fmacro and accuracy. RF works better with lemma unigrams and bigrams, whereas SVC works better with the full set of parameters and Quita+Lemma + POS 1-, 2-grams set.

Algorithms perform slightly better on female texts than on male texts. It should also be noted that adding new features to lemma unigrams does not give a crucial increase in F-macro/Accuracy in case of SVC and leads to lower results in case of RF.

As our results indicate, regression with the full set of parameters gives slightly better results than with Lemma 1-, 2-grams, which is similar to our observations in the classification case. Similar to findings reported in Nquyen et al. (2011), our results indicate that using only POS is not an effective strategy in age regression.

We present the distribution of true and predicted results using regression in Fig. 2.

We also performed regression analysis on male and female texts separately with the full feature set: QUITA + lemma 1-, 2-grams +pos 1-, 2-grams + tense, aspect (alpha = 0.003), and have found out that regression also works better for female texts in terms of both MAE and R2 (see Table 4).

Table 1. Results for classification with RandomForest, n = 500 (F-macro/accuracy)

Setting	Lemma unigrams	Lemma 1-, 2-grams	Lemma 1-, 2-grams + POS 1,2-grams	QUITA + Lemma 1-, 2-grams + POS 1-, 2-grams	QUITA + tense, aspect + Lemma 1-, 2-grams + POS 1-, 2-grams
W/O gender	0.43/0.432	**0.492/0.495**	0.462/0.468	0.459/0.464	0.458/0.462
With gender	0.431/0.433	0.49/0.493	0.463/0.468	0.461/0.466	0.446/0.452
Male		0.503/0.504			0.476/0.479
Female		**0.506/0.513**			0.474/0.482

In both regression and classification tasks, we have obtained the results above the baseline. It should be noted that the addressed task itself is rather difficult, as it was shown by, for example, Nguyen et al. (2014). According to the authors, predicting older users (especially older than 30 years) is harder than predicting young ones – both for humans and for machines.

Table 2. Results for classification with LinearSVC, pen = l1, C = 6 (F-macro/accuracy)

Setting	Lemma unigrams	Lemma 1-, 2-grams	Lemma 1-, 2-grams + POS 1-, 2-grams	QUITA + Lemma 1-, 2-grams + POS 1-, 2-grams	QUITA + tense, aspect + Lemma 1-, 2-grams + POS 1-, 2-grams
W/O gender	0.434/0.437	0.478/0.483	0.484/0.49	0.489/0.494	0.483/0.489
With gender	0.434/0.438	0.476/ 0.481	0.489/0.495	**0.492/0.498**	**0.492/0.498**
Male				0.47/0.472	0.472/0.475
Female				0.503/ 0.511	**0.508/0.515**

Table 3. Results for regression

Lasso, alpha = 0.0015	N features selected	MAE/R2
lemma+pos 1-, 2-grams + quita	990	5.99/0.12
lemma+pos 1-, 2-grams	981	5.99/0.12
Lemma 1-, 2-grams	882	6.17/0.09
Pos 1-, 2-grams	232	R2 < 0.02
lemma+pos 1-, 2-grams + quita + tense, aspect	997	**5.98/0.12**

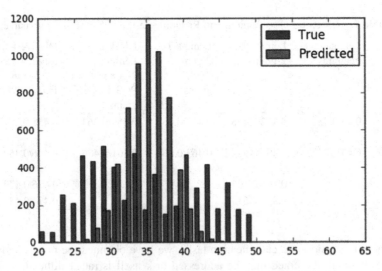

Fig. 2. True and predicted results using regression

Table 4. Results of lasso regression for male and female texts

Authors	N features selected	MAE/R2	Baseline (dummy mean prediction) MAE
Female	822	**5.70/0.18**	6.59
Male	863	6.14/0.09	6.65

4.2 Feature Analysis

Our feature analysis revealed that most features are best at distinguishing between classes 20–30 and 40–50.

The most important finding consistent trough all experiments is that older people (40–50) tend to use less pronoun *я* (I), whereas they use more nouns as well as noun_noun bigrams than younger bloggers. Older people also use more adj_noun bigrams and adjectives overall. Older people also tend to use less infinitives and adverbs as well as comparative forms. The value of Token Length Frequency Spectrum (one of QUITA parameters) is higher in texts of older people.

Younger people tend to use more I-forms as well as forms *мой* (my) and more verbs and conjunctions overall, as well as bigrams, *_но* (, but), *но_я* (but I), *я_весь* (I am all), *как_я* (like me), more impersonal forms like *хотеться* (I want), *кажется* (it seems that), more intensifiers (*очень* very).

Older people use more lemma *поскольку* (since), *народ* (folks), *русский* (Russian), *наш* (ours), more citations. Note that authors in Nguyen et al. (2013) also report that older people have a higher usage of links and hashtags, which can be associated with information sharing.

To sum up, the following patterns are revealed. In general, younger people use more egocentric and emotional words, whereas with age people tend to become more "formal" (informative) in their writing (Argamon et al. 2007). Our results are consistent with findings by Pennebaker and Stone (2003) who found that as people get older, they tend to use fewer negations and make fewer self-references as well as with results of Twitter study (Nguyen et al. 2013) which revealed that younger persons talk more about themselves. Younger people preference of features that indicate stance and emotional involvement (intensifiers, emoticons) was shown in many studies (see Nguyen 2017, for review).

5   Conclusions

We have performed age prediction experiments on the collected corpus of Russian-language texts from LiveJournal, which is the first attempt to identify age based on Russian texts. We have used different approaches to age prediction based on classification models as well as on regression models. Unlike many other studies, we did not make any gaps between age classes, to make the task as realistic as possible. In addition, we analyzed texts written by adults (20–50), i.e. we did not include in our corpus texts by adolescents and aged people, which made our task more difficult than the ones approached in previous work. We have obtained the results above the baseline both in classification (Acc $\sim$0.5 VS 0.33) and in regression experiments (MAE $\sim$6.0 VS 6.6, R2 = 0.13). The results are rather modest, but comparable to the results of similar tasks in other languages.

We have found out that very basic features such as unigrams and bigrams of lemmas work well, and adding additional features (POS, lexical diversity) gives a slight or no improvement in accuracy (depending on the experiment settings). We have also revealed that both classification and regression models work better for female texts than for male texts. Contrary to our expectations, however, adding gender as feature did not significantly improve accuracies of the models.

As part of our future work, we are planning to test new features, namely LIWC-based parameters, to build new corpora and to examine the cross-genre effect in age prediction, which corresponds to the latest trend in authorship profiling area.

Acknowledgment. Funding of the project "Identifying the Gender and Age of Online Chatters Using Formal Parameters of their Texts" from the Russian Science Foundation (no. 16-18-10050) is gratefully acknowledged.

References

Alekseev, A., Nikolenko, S.I.: Predicting the age of social network users from user-generated texts with word embeddings. In: Proceedings of the AINL FRUCT 2016 Conference, pp. 1–11. IEEE, St. Petersburg (2017)

Argamon, S., Koppel, M., Pennebaker, J.W., Schler, J.: Mining the blogosphere: age, gender and the varieties of self-expression. First Monday **12**(9) (2007). http://firstmonday.org/ojs/index.php/fm/article/view/2003/1878

Gomzin, A., Laguta, A., Stroev, V., Turdakov, D.: Detection of author's educational level and age based on comments analysis. Paper presented at Dialogue 2018, Moscow, 30 May–2 June 2018. http://www.dialog-21.ru/media/4279/gomzin_turdakov.pdf (2018)

Jones, E., Oliphant, T., Peterson, P.: SciPy: open source scientific tools for Python (2014). https://www.scipy.org/. Accessed 21 May 2018

Korobov, M.: Morphological analyzer and generator for Russian and Ukrainian languages. In: Khachay, M.Y., Konstantinova, N., Panchenko, A., Ignatov, D.I., Labunets, V.G. (eds.) AIST 2015. CCIS, vol. 542, pp. 320–332. Springer, Cham (2015). https://doi.org/10.1007/978-3-319-26123-2_31

Kubát, M., Matlach, V., Čech, R.: Studies in Quantitative Linguistics 18: QUITA-Quantitative Index Text Analyzer. RAM-Verlag, Lüdenscheid (2014)

Litvinova, T., Rangel, F., Rosso, P., Seredin, P., Litvinova, O.: Overview of the RusProfiling PAN at FIRE track on cross-genre gender identification in Russian. In: CEUR Workshop Proceedings, pp. 1–7 (2017)

Litvinova, T., Seredin, P., Litvinova, O., Zagorovskaya, O.: Identification of gender of the author of a written text using topic-independent features. Pertanika J. Soc. Sci. Hum. **26**(1), 103–112 (2018)

Litvinova, T., Seredin, P., Litvinova, O., Zagorovskaya, O.: Profiling a set of personality traits of text author: what our words reveal about us. Res. Lang. **14**(4), 409–422 (2016)

Lui, M., Baldwin, T.: langid.py: an off-the-shelf language identification tool. In: Proceedings of the ACL 2012 System Demonstrations, pp. 25–30 (2012)

Nguyen, D., Dogruöz, A.S., Rosé, C.P., de Jong, F.: Computational sociolinguistics: a survey. Comput. Linguist. **42**(3), 537–593 (2016)

Nguyen, D., Gravel, R., Trieschnigg, D., Meder, T.: How old do you think I am? A study of language and age in Twitter. In: Proceedings of the Seventh International AAAI Conference on Weblogs and Social Media, pp. 439–448. Boston, Massachusetts, USA (2013)

Nguyen, D., Smith, N.A., Rosé, C.P.: Author age prediction from text using linear regression. In: Proceedings of the 5th ACL-HLT Workshop on Language Technology for Cultural Heritage, Social Sciences, and Humanities, pp. 115–123. Association for Computational Linguistics (2011)

Nguyen, D., et al.: Why gender and age prediction from tweets is hard: lessons from a crowdsourcing experiment. In: Proceedings of COLING 2014, the 25th International Conference on Computational Linguistics: Technical Papers, Dublin, Ireland, pp. 1950–1961 (2014)

Nguyen, D.: Text as social and cultural data: a computational perspective on variation in text. Ph. D. dissertation, University of Twente (2017)

Panicheva, P., Mirzagitova, A., Ledovaya, Y.: Semantic feature aggregation for gender identification in Russian Facebook. In: Filchenkov, A., Pivovarova, L., Žižka, J. (eds.) AINL 2017. CCIS, vol. 789, pp. 3–15. Springer, Cham (2018). https://doi.org/10.1007/978-3-319-71746-3_1

Pedregosa, F., et al.: Scikit-learn: machine learning in Python. J. Mach. Learn. Res. **12**, 2825–2830 (2011)

Pennebaker, J.W., Francis, M.E., Booth, R.J.: Linguistic Inquiry and Word Count: LIWC 2001. Lawrence Erlbaum, Mahwah (2001)

Pennebaker, J.W., Stone, L.D.: Words of wisdom: language use over the life span. J. Personal. Soc. Psychol. **85**(2), 291–301 (2003)

Rangel, F., Rosso, P., Verhoeven, B., Daelemans, W., Potthast, M., Stein, B.: Overview of the 4th author profiling task at PAN 2016: cross-genre evaluations. In: Balog, K., et al. (eds.) Working Notes Papers of the CLEF 2016 Evaluation Labs. CEUR Workshop Proceedings, pp. 750–784 (2016)

Rosenthal, S., McKeown, K.: Age prediction in blogs: a study of style, content, and online behavior in pre- and post-social media generations. In: Proceedings of the 49th Annual Meeting of the Association for Computational Linguistics: Human Language Technologies, vol. 1, pp. 763–772 (2011)

Sboev, A., Litvinova, T., Gudovskikh, D., Rybka, R., Moloshnikov, I.: Machine learning models of text categorization by author gender using topic-independent features. Procedia Comput. Sci. **101**, 135–142 (2016)

Sboev, A., Moloshnikov, I., Gudovskikh, D., Selivanov, A., Rybka, R., Litvinova, T.: Automatic gender identification of author of Russian text by machine learning and neural net algorithms in case of gender deception. Procedia Comput. Sci. **123**, 417–423 (2018)

Schler, J., Koppel, M., Argamon, S., Pennebaker, J.W.: Effects of age and gender on blogging. In: Proceedings of AAAI Spring Symposium on Computational Approaches to Analyzing Weblogs, pp. 199–205. Menlo Park, California (2006)

Tutubalina, E., Nikolenko, S.: Automated prediction of demographic information from medical user reviews. In: Prasath, R., Gelbukh, A. (eds.) MIKE 2016. LNCS (LNAI), vol. 10089, pp. 174–184. Springer, Cham (2017). https://doi.org/10.1007/978-3-319-58130-9_17

Stierlitz Meets SVM: Humor Detection in Russian

Anton Ermilov[1]([✉]), Natasha Murashkina[1], Valeria Goryacheva[2], and Pavel Braslavski[1,3,4]

[1] National Research University Higher School of Economics, Saint Petersburg, Russia
anton.yermilov@gmail.com, murnatty@gmail.com, pbras@yandex.ru
[2] ITMO University, Saint Petersburg, Russia
gor.ler177@gmail.com
[3] Ural Federal University, Yekaterinburg, Russia
[4] JetBrains Research, Saint Petersburg, Russia

Abstract. In this paper, we investigate the problem of the humor detection for Russian language. For experiments, we used a large collection of jokes from social media and a contrast collection of non-funny sentences, as well as a small collection of puns. We implemented a large set of features and trained several SVM classifiers. The results are promising and establish a baseline for further research in this direction.

Keywords: Humor recognition · Evaluation

1 Introduction

Humor is an important aspect of human communication. Rapid proliferation of conversational agents, voice interfaces, and chatbots, as well as the need to analyze large volumes of social media texts make the task of humor detection highly relevant.

In this study, we used a subset of an existing collection of short jokes in Russian from social media and also collected a contrast collection of non-funny sentences. In addition, we collected a small collection of puns to test the developed method on this special kind of humorous content. We engineered a wide range of features that reflects different aspects of language – lexical, semantic, structural, etc. We trained several binary classifiers and evaluated contribution of individual feature groups to the classification quality. The obtained results demonstrate acceptable performance and provide the basis for further research in this direction. To the best of our knowledge, current study is the first experiment on automatic detection of humor in the Russian language.

Stierlitz is a Soviet spy working deep undercover in Nazi Germany, a protagonist of a TV series from 1972 based on a novel by Yulian Semionov. Stierlitz became a popular joke character in Soviet and post-Soviet culture.

D. Ustalov et al. (Eds.): AINL 2018, CCIS 930, pp. 178–184, 2018.
https://doi.org/10.1007/978-3-030-01204-5_17

2   Related Work

The humor recognition is usually formulated as a classification task with a wide variety of features – syntactic parsing, alliteration and rhyme, antonymy and other WordNet relations, dictionaries of slang and sexually explicit words, polarity and subjectivity lexicons, distances between words in terms of *word2vec* representations, etc. In their pioneering work, Michalcea and Strapparava [7] compiled a dataset of humorous and non-humorous sentences in English – 16,000 one-line jokes from the web and 16,000 sentences from the news, the British National Corpus, collections of proverbs, as well as collection of common sense sentences and performed a classification experiment with different features. A follow-up study [6] investigated humor features in more detail. Zhang and Liu [14] experimented with the humor detection in tweets. Yang et al. [13] introduced the notion of humor anchors – words and phrases 'responsible' for a humorous effect, experimented with a large collection of puns and explored a wide range of features for the humor detection, including those based on vector representations. Shahaf et al. [12] addressed the task of ranking cartoon captions provided by the readers of New Yorker magazine. They employed a wide range of linguistic features as well as features from manually crafted textual descriptions of the cartoons. Two recent shared tasks dealing with humor within the SemEval campaign signal a growing interest in the topic [8,9]. A cognate task is detection of other forms of figurative language such as irony and sarcasm [10,11].

3   Data

In the current study we used a collection of jokes in Russian from online social networks that we obtained from the authors of [2]. The collection consists of about 63,000 one-liners collected from VK and Twitter. The jokes are in plain text, i.e. media content, URLs, and hashtags are removed; more details about the dataset can be found in the paper. From this collection, we randomly sampled 47,000 items for our experiments. To build a contrast collection, we gathered sentences from Russian classical novels (28,000), news headlines (13,000) and proverbs (6,000). We did not make efforts to ensure lexical similarity of the funny and non-funny parts of the collection, as the authors of [7] did. The only additional parameter was the length – sentences of 25 words and shorter are included in the collection (average length is 14 words). For experiments, the collection was splitted into training/test sample in a ratio of 80/20.

In addition, we manually created a small collection of puns. In total, there are 200 jokes with a word play in the collection, most of them are associated with the "Omsk Ptitsa" meme and the Stierlitz jokes. We used this collection only for testing classifiers trained on the data from the BIG collection.

4   Features

Based on literature review and manual inspection of the collection, we implemented six groups of text features that can potentially distinguish between humorous and non-humorous content. The features are briefly described below.

Bag-of-Words (BOW). Each text is presented as a 12,000-dimensional binary vector. The intuition behind the feature is that some words are quite specific for the humorous content.

Sentence2Vec (S2V) is aimed at capturing sense of the text as a 300-dimensional vector. We summed up vectors of individual words in the text weighed by their IDFs. We used pre-trained word2vec vectors available through the RusVectōrēs project [5]. IDF weights are calculated using the Russian National Corpus data.[1]

Structural features (SF) are shallow features capturing the complexity of the text (average word length in characters and syllables, fraction of stopwords) and its organization – punctuation marks, question words and certain conjunctions.

Lexical Features (LF). This group of word-level features includes:

- minimum/maximum word frequencies calculated using RNC statistics;
- a share of words with non-common usage labels (*informal, offensive, vulgar,* etc.) from the Russian Wiktionary[2];
- a maximum number of possible POS tags over all words and a proportion of nouns/verbs/adjectives/numerals in the text based on the PyMorphy output [4];
- a presence of proper names and parenthetical words.

RuWordNet features (RWN). Using the RuWordNet thesaurus[3] we calculated the following features:

1. Ambiguity
   - a *sense combination*, formalized as $\sum \log(n_{w_i})$, where n_{w_i} is the number of senses of the word w_i (we account only for nouns, verbs and adjectives present in the RuWordNet);
   - the largest *path similarity* over all word-sense pairs, whereas the *path similarity* is the minimal distance between word-senses in thesaurus graph (lower values correspond to semantically closer senses);
2. Domains
   - a number of different domains associated with words in the text;
   - a number of words that belong to different domains.
3. Number of synonym and antonym pairs in the text.

---

[1] http://ruscorpora.ru/corpora-freq.html.
[2] https://ru.wiktionary.org/.
[3] http://ruwordnet.ru/.

Word2Vec (W2V). Following [13], we calculate two word2vec-based features:

- *disconnection*: the maximum semantic distance of word pairs in a sentence;
- *repetition*: the minimum semantic distance of word pairs in a sentence.

5   Results and Discussion

We used the LibSVM [3] to train classifiers. We experimented with various combinations of feature groups. The Table 1 below summarizes results. The reported figures correspond to the linear SVM that delivered better results in our experiments than SVMs with polynomial and RBF kernels. Columns 2–5 report results achieved on the test set of the 'big' dataset of one-liners and non-funny sentences; precision, recall, and F1 correspond to the humorous class. The last column of the Table reports recall of the classifier trained on the training set from the 'big' dataset and then applied to the small collection of puns.

As can be seen from the Table below, the classification based solely on bag-of-words features is a very strong baseline ($F1 = 0.846$ on the BIG dataset, $R = 0.671$ on the PUNS). On the one hand, it can be explained through the way the collection was built: positive and negative classes are quite distinctive on the lexical level. On the other hand, recall on the independent PUNS collection is also relative high. $S2V$ is a runner-up among individual feature groups ($F1 = 0.811$ on the BIG dataset, $R = 0.601$ on the PUNS). Thus, S2V shows no generalization over individual words. We can hypothesize that vector representation 'flattens' the **sentence** meaning and doesn't account for possible alternative interpretation, which might be crucial for the humorous content. The combination of these two sentence meaning representations (BOW + S2V) improves over both approaches and achieves the best score on the PUNS collection (*recall* = 0.695). Other feature groups, taken separately, demonstrate much lower performance.

The combination of BOW with features, potentially reflecting semantic relations between words in the sentence (RWN and W2V), delivers mixed results. Adding RWN features improves precision on the humorous class ($P = 0.863$), while W2V degrades overall results on the 'big' collection. One can argue that manually crafted semantic resources are still a viable alternative for general-purpose semantic representations based on neural networks, especially for high-precision results. However, these combinations behave reversely on the PUNS collection. BOW + W2V shows second-best result on the PUNS ($R = 0.676$). Results in the Table 1 support in general the claim that more features mean the better classification quality. The combination of all features delivers best results on the BIG dataset ($F1 = 0.884$). However, the addition of two W2V features has a marginal impact. These results somewhat contradict the feature importance considerations reported in [13]. However, a direct comparison between different datasets in different languages is hardly possible.

A manual inspection of misclassified jokes reveals that the majority of them are unfunny according to our subjective opinion. For example, this item from the jokes collection looks rather like a proverb:
Хочешь идти быстро — иди один. Хочешь уйти далеко — идите вместе.

Table 1. Humor recognition results.

Feature set	BIG				PUNS
	Accuracy	Precision	Recall	F1	Recall
BOW	0.848	0.855	0.837	0.846	0.671
S2V	0.813	0.820	0.801	0.811	0.601
SF	0.671	0.658	0.715	0.685	0.385
LF	0.618	0.603	0.690	0.643	0.117
RWN	0.563	0.626	0.311	0.416	0.211
W2V	0.527	0.579	0.196	0.293	0.160
BOW + RWN	0.850	0.863	0.832	0.847	0.638
BOW + LF	0.850	0.856	0.842	0.849	0.559
BOW + SF	0.869	0.872	0.863	0.868	0.568
BOW + W2V	0.846	0.853	0.836	0.845	0.676
BOW + SF + LF + RWN	0.871	0.873	0.862	0.870	0.521
S2V + RWN	0.814	0.824	0.798	0.811	0.592
S2V + LF	0.818	0.826	0.806	0.815	0.526
S2V + SF	0.839	0.848	0.826	0.837	0.498
S2V + W2V	0.814	0.822	0.802	0.812	0.606
S2V + SF + LF + RWN	0.846	0.854	0.834	0.844	0.521
BOW + S2V	0.868	0.873	0.861	0.867	**0.695**
BOW + S2V + SF + LF + RWN	**0.885**	**0.892**	**0.875**	**0.884**	0.620
BOW + S2V + SF + LF + RWN + W2V	**0.885**	**0.892**	**0.876**	**0.884**	0.615

If you want to go fast, go alone. If you want to go far, go together.

Other false negatives are *referential* jokes that require some world knowledge to comprehend them (see [1] for details). For example, this joke refers to dung beetles rolling balls out of dirt and ball-shaped Raffaello candy:

Жук-навозник на День рождения прикатил жене рафаэлку.

A dung beetle brought his wife a Raffaello as a birthday present.

Considering puns, we hypothesize that the following joke was not recognized because of a very scarce context (the pun plays around two senses of the verb *звонить* – to ring/to phone).

Звонил колокол. Угрожал. // *The bell rang. Threatened.*

Most false positives are literature excerpts, for example:

Все подняли головы, прислушались, и из леса, в яркий свет костра, выступили две, держащиеся друг за друга, человеческие, странно одетые фигуры. // *Everyone lifted their heads, listening closely, and two strangely dressed human figures stood out from the forest into the bright light of the fire, holding each other.*

Many incorrectly classified excerpts were rather long. Possibly, many word combinations result in triggering some semantic features. Moreover, sentences from fiction works may contain some figurative language.

6   Conclusion

We prepared data and conducted experiments aimed at the humor detection in short Russian texts. We implemented a wide range of text features and conducted a comparative study of their impact on the classification quality. The obtained results form a strong baseline for future research in the field of a computational humor on Russian language data. Pun collection used in the study is freely available for research.[4] In the future, we plan to employ a more elaborate sampling of negative (non-humorous) examples. In addition, we plan to develop methods and features that better capture a word play; expand the collection of puns and conduct a finer-grained annotation of jokes. In the framework of this study, we haven't investigated several features potentially useful for the humor detection: phonetic and syntactic features, as well as those based on sentiment lexicons. We plan to address these tasks in the future.

Acknowledgments. We thank Valeria Bolotova and Vladislav Blinov for sharing their humor dataset, as well as Natalia Loukachevitch for providing us with the RuWordNet data.

References

1. Attardo, S.: Linguistic Theories of Humor. Mouton de Gruyter, Berlin (1994)
2. Bolotova, V., et al.: Which IR model has a better sense of humor? Search over a large collection of jokes. In: Dialogue, pp. 29–42 (2017)
3. Chang, C.C., Lin, C.J.: LIBSVM: a library for support vector machines. ACM Trans. Intell. Syst. Technol. **2**, 27:1–27:27 (2011)
4. Korobov, M.: Morphological analyzer and generator for russian and ukrainian languages. In: Khachay, M.Y., Konstantinova, N., Panchenko, A., Ignatov, D.I., Labunets, V.G. (eds.) AIST 2015. CCIS, vol. 542, pp. 320–332. Springer, Cham (2015). https://doi.org/10.1007/978-3-319-26123-2_31
5. Kutuzov, A., Kuzmenko, E.: WebVectors: a toolkit for building web interfaces for vector semantic models. In: Ignatov, D.I., et al. (eds.) AIST 2016. CCIS, vol. 661, pp. 155–161. Springer, Cham (2017). https://doi.org/10.1007/978-3-319-52920-2_15
6. Mihalcea, R., Pulman, S.: Characterizing humour: an exploration of features in humorous texts. In: Gelbukh, A. (ed.) CICLing 2007. LNCS, vol. 4394, pp. 337–347. Springer, Heidelberg (2007). https://doi.org/10.1007/978-3-540-70939-8_30
7. Mihalcea, R., Strapparava, C.: Learning to laugh (automatically): computational models for humor recognition. Comput. Intell. **22**(2), 126–142 (2006)
8. Miller, T., Hempelmann, C., Gurevych, I.: SemEval-2017 Task 7: detection and interpretation of English puns. In: SemEval (2017)
9. Potash, P., Romanov, A., Rumshisky, A.: SemEval-2017 Task 6: #HashtagWars: learning a sense of humor. In: SemEval, pp. 49–57 (2017)
10. Rajadesingan, A., Zafarani, R., Liu, H.: Sarcasm detection on Twitter: a behavioral modeling approach. In: Proceedings of WSDM, pp. 97–106 (2015)

---

[4] http://eranik.me/humor-detection.

11. Reyes, A., Rosso, P., Veale, T.: A multidimensional approach for detecting irony in Twitter. Language resources and evaluation **47**(1), 239–268 (2013)
12. Shahaf, D., Horvitz, E., Mankoff, R.: Inside jokes: identifying humorous cartoon captions. In: Proceedings of KDD, pp. 1065–1074 (2015)
13. Yang, D., Lavie, A., Dyer, C., Hovy, E.: Humor recognition and humor anchor extraction. In: Proceedings of EMNLP, pp. 2367–2376 (2015)
14. Zhang, R., Liu, N.: Recognizing humor on Twitter. In: CIKM, pp. 889–898 (2014)

Interactive Attention Network
for Adverse Drug Reaction Classification

Ilseyar Alimova$^{(\boxtimes)}$ and Valery Solovyev

Kazan (Volga Region) Federal University, Kazan, Russia
alimovailseyar@gmail.com, maki.solovyev@mail.ru

Abstract. Detection of new adverse drug reactions is intended to both improve the quality of medications and drug reprofiling. Social media and electronic clinical reports are becoming increasingly popular as a source for obtaining the health-related information, such as identification of adverse drug reactions. One of the tasks of extracting adverse drug reactions from social media is the classification of entities that describe the state of health. In this paper, we investigate the applicability of Interactive Attention Network for identification of adverse drug reactions from user reviews. We formulate this problem as a binary classification task. We show the effectiveness of this method on a number of publicly available corpora.

Keywords: Adverse drug reactions · Text mining
Natural language processing · Health social media analytics
Machine learning · Deep learning

1 Introduction

The rapid development of social media and digital collections of scientific publications intended an increase in the volume of unstructured information represented by texts in natural languages. Recent research has demonstrated that unstructured texts are a promising source for problems in the medical area, in particular, the tasks of computerized clinical decision support [12], extracting medical problems from electronic clinical documents [34], predicting future disease risk [59], massive deterioration of well-being and detecting new adverse drug reactions [13,44]. Extracting of adverse drug reactions (ADRs) in post-marketing period is becoming increasingly popular, as evidenced by the growth of ADR monitoring systems [16,53,54]. Due to the various limitations of pre-approval clinical trials, it is not possible to detect all the consequences of using a particular drug before sending it for sale. The importance of identifying new ADRs is due to the fact that adverse drug reactions are a significant cause of morbidity and mortality, often identified only post-marketing [10,25,43].

Information about adverse drug reactions can be found in the texts of social media, health-related forums, and electronic health records. This amount of information cannot be processed manually, therefore, methods based on natural

© Springer Nature Switzerland AG 2018
D. Ustalov et al. (Eds.): AINL 2018, CCIS 930, pp. 185–196, 2018.
https://doi.org/10.1007/978-3-030-01204-5_18

language processing are actively developed [15,49]. This is a challenging task as there are multiple types of drug-related discussions confounded with patient adverse drug event reports. Thus at the first step, all information related to a state of health are extracting using named entity recognition systems. Then all obtained entities are classified in order to distinguish ADRs from indication and patient history. In this article, we focused on the task of binary classification.

Our approach is based on the idea that sentiment analysis of the text can help with adverse drug reaction classification [23,40]. Existing works utilized sentiment lexicons, negations or general evaluation of sentiment of the entity context. More recently, deep learning models with attention have become very popular for sentiment analysis [18,28,30,60]. Interactive attention networks have shown promising results in various NLP tasks including machine translation [33], question answering systems [26,57], document classification [61]. Ma et al. proposed the interactive attention networks (IAN) for aspect-level sentiment classification [30]. The main idea of this approach is to learn own representations for targets and contexts via interactive learning. Previous studies ignored the separate modeling of targets. Experiments on SemEval 2014 dataset [31] demonstrate that IAN achieves the state-of-the-art performance. In this paper, we investigate the effectiveness of the IAN for ADR classification. We conduct extensive experiments on five real-life corpora from Askapatient.com, Twitter.com, PubMed and demonstrate the efficiency of the proposed approach over a strong baseline for classification based on machine learning with hand-crafted features.

2   Related Work

Different approaches are utilized to identify adverse drug reactions. First works were limited in the number of study drugs and targeted ADRs due to limitations of traditional lexicon-based approaches [5,29]. In order to eliminate these shortcomings, rule-based methods have been proposed [37,38]. These methods capture the underlying syntactic and semantic patterns from social media posts. Third group of works utilized popular machine learning models, such as support vector machine (SVM) [1,6,29,40,49], conditional random fields (CRF) [3,35], and random forest (RF) [45]. The most popular hand-crafted features are n-grams, parts of speech tags, semantic types from the Unified Medical Language System (UMLS), the number of negated contexts, the belonging lexicon based features for ADRs, drug names, and word embeddings [11]. One of the tracks of recently held competition "Social Media Mining for Health Applications Shared Task 2016" was devoted to ADR classification on a tweet level. Best performance is achieved by SVM classifiers with a variety of surface-form, sentiment, and domain-specific features [22]. This classifier obtained 43.5% F-measure for 'ADR' class. However, Sarker and Gonsales outperformed these result utilizing SVM with a more rich set of features and the tuning of the model parameters and obtained 53.8% F-measure for 'ADR' class [50]. However, these results are still behind the current state-of-the-art for general text classification [24].

Modern approaches for the extracting of ADRs are based on neural networks. Saldana adopted CNN for the detection of ADR relevant sentences [36]. Huynh et al. applied convolutional recurrent neural network (CRNN), obtained by concatenating CNN with a recurrent neural network (RNN) and CNN with the additional weights [19]. Gupta et al. utilized a semi-supervised method based on co-training [14]. Chowdhury et al. proposed a multi-task neural network framework that in addition to ADR classification learns extract ADR mentions [9].

While recent research has been devoted to automatic analysis of biomedical texts written in English, little has been done to analyze other languages. There are studies for Spanish texts [42] and French records [41]. In the industry field, there is also a company Web-Radr[1], which extracts adverse drug reactions from user feedback in social networks for 7 languages: Danish, German, Croatian, English, Spanish, French and Portuguese.

Methods for sentiment analysis are actively adopted in the medical domain as well as in other domains [20,47,52,56]. In [7,46,51] Biyani et al. and Rodrigues et al. focused on a classification of records from the social network of people with cancer. Sokolova et al. tested several classifiers to evaluate the tonality of tweets related to medicine [55]. Salas-Zárate et al. proposed method to sentiment analysis of tweets associated with diabetes [48]. Study [32] was aimed at finding various emotions in medical texts: joy, anger, surprise, etc. These studies are necessary for doctors in order to make decisions about the patient's treatment. Cambria et al. presented the Sentic PROMs system, in which sentiment analysis was applied for a general evaluation of the quality of healthcare [8].

To sum up this section, we note that there has been little work on utilizing neural networks for ADR classification task. Most of the works used classical machine learning models, which are limited to linear models and manual feature engineering [1,3,6,29,35,40,45,49]. Most methods for extracting ADR so far dealt with extracting information from the mention itself and a small window of words on the left and on the right as a context, ignoring the broader context of the text document where it occurred [1,3,6,11,23]. Finally, in most of the works experiments were conducted on a single corpus.

Our work differs from the mentioned works in several important aspects. First, we experiment with IAN model. Second, we use as a contest the whole sentence, in which the mention of ADRs occurs. Third, we train a neural network on a combination of datasets and evaluate the qualitative gain of the proposed model. Fourth, we use word embeddings trained on texts about health from social media.

3   Corpora

We conducted our experiments on four corpora: CADEC, Twitter, MADE, TwiMed. Further, we briefly describe each dataset.

CADEC. CSIRO Adverse Drug Event Corpus (CADEC) consists of annotated user reviews written about Diclofenac or Lipitor on askapatient.com [21]. There are five types of annotations: 'Drug', 'Adverse effect', 'Disease', 'Symptom', and

---

[1] https://web-radr.eu/.

Table 1. Summary statistics of corpora.

Corpus	Documents	ADR	non-ADR	Max sentence length
CADEC [21]	1231	5770	550	236
MADE [27]	876	1506	37077	173
TwiMed-Pubmed [2]	1000	264	983	150
TwiMed-Twitter [2]	637	329	308	42
Twitter [39]	645	569	76	37

'Finding'. We grouped diseases, symptoms, and findings as a single class called 'non-ADR'.

MADE. MADE corpus consists of de-identified electronic health record notes from 21 cancer patients [27]. The corpus is developed specially for the NLP challenges for Detecting Medication and Adverse Drug Events competition in 2017. Each record annotated with medications and relations to their corresponding attributes, indications and adverse events. We grouped annotations corresponding to the diseases in class 'non-ADR', such as 'Indication' and 'SSLIF'.

TwiMed. TwiMed corpus consists of sentences extracted from PubMed and tweets. This corpus contains annotations of diseases, symptoms, and drugs, and their relations. If the relationship between disease and drug was labeled as 'Outcome-negative', we marked disease as ADR, otherwise, we annotate it as 'non-ADR' [2].

Twitter. Twitter corpus include tweets about drugs. There are three annotations: 'ADR', 'Indication' and 'Other'. We consider 'Indication' and 'Other' as 'non-ADR' [39]. Due to the Twitter copyright concerns, the Twitter datasets consists of identifiers of the user and the tweet using these identifiers, tweets can be found. Therefore, tweets can be downloaded via their IDs. A number of tweets become unavailable at the time of preparing this article. For this reason, we were able to use only part of the dataset with the surviving texts of tweets.

Summary statistics of corpora are presented in Table 1. As shown in this table, CADEC and MADE corpora contain a much larger number of annotations than TwiMed and Twitter.

4   Interactive Attention Network

Supervised models are facing three important challenges in aspect level sentiment classification. The first challenge is to represent the context of a target (ADR and non-ADR in our study). The second challenge is to generate a target representation, which can interact with its context. The third challenge is to identify the important sentiment words for the target. Let us take "He was unable to sleep last night because of pain" as an example. In this case, due to 'unable to sleep' is an effect of pain and marked as non-ADR. However, in

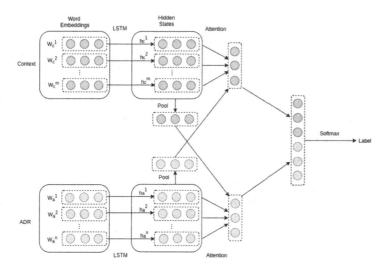

Fig. 1. The overall architecture of IAN.

the sentence "Became unable to walk without a cane, unable to sleep, kidney problems (urine like root beer)" the target 'unable to sleep' is ADR.

In this work, we utilize Interactive Attention Network (IAN) [30]. In order to address first two challenges, IAN learns representations for targets and contexts. Basically, this model is composed of two parts which model the target and context interactively. Using word embeddings as input, LSTM layers are employed to obtain hidden states of words for a target and its context, respectively. The average value of the target's hidden states and the context's hidden states are used to calculate attention vectors. IAN uses attention mechanisms to detect the important words of the target expression and its full context. After computing the attention weights, IAN compute context and target representations c_r and t_r based on the attention vectors. These representations are concatenated into a single vector d for a classifier. The overall architecture of IAN model is shown in Fig. 1. Please refer to [30] for the details of IAN.

5   Experiments

In this section, we describe our experiments with IAN.

5.1   Baseline Method

We compare our approach with a feature-rich classifier from [1]. This method is based on SVM with Linear kernel. A set of experiments showed that unigrams and bigrams, part of speech tags, sentiment, cluster-based representation and semantic types from Unified Medical Language System features are the most effective to classify ADRs. The part of speech feature consist of the number of nouns, verbs, adverbs and adjectives. For sentiment feature, the following lexicons were used: SentiWordNet [4], MPQA Subjectivity Lexicon [58], Bing Liu's dictionaries [17]. The cluster-based representation feature utilized clusters from [35] with Brown hierarchical clustering algorithm. The last feature consist of the number of tokens from each UMLS semantic types. The classifier was compared with CNN and SVM model with another set of features. The effectiveness of the methods was evaluated on the CADEC [21] and Twitter [50] corpora.

5.2   Result and Analysis

We used vector representation trained on social media posts from [35]. Word embedding vectors were obtained with using word2vec trained on unlabeled Health corpus consists of 2.5 million reviews written in English. We used an embedding size of 200, local context length of 10, the negative sampling of 5,

Table 2. Classification results of the compared methods for CADEC corpus.

Method	ADR			non-ADR			Macro		
	P	R	F	P	R	F	P	R	F
Feature-rich SVM	.964	.969	.967 ± .004	.659	.620	.638 ± .018	.811	.795	.802 ± .010
IAN	**.966**	**.972**	**.969 ± .005**	**.699**	**.637**	**.662 ± .018**	**.832**	**.805**	**.815 ± .011**
IAN (all)	.962	.924	.943 ± .005	.437	.615	.508 ± .036	.700	.770	.726 ± .018

Table 3. Classification results of the compared methods for Twitter corpus.

Method	ADR			non-ADR			Macro		
	P	R	F	P	R	F	P	R	F
Feature-rich SVM	.937	.952	.944 ± .014	.602	.520	.554 ± .014	.769	.736	.749 ± .104
IAN	**.951**	**.957**	**.954 ± .010**	**.654**	**.627**	**.634 ± .114**	**.802**	**.792**	**.794 ± .062**
IAN (all)	.935	.838	.883 ± .026	.320	.560	.404 ± .096	.627	.699	.643 ± .058

Table 4. Classification results of the compared methods for MADE corpus.

Method	ADR			non-ADR			Macro		
	P	R	F	P	R	F	P	R	F
Feature-rich SVM	.551	**.582**	.562 ± .093	**.984**	.981	.982 ± .001	.767	.782	.772 ± .046
IAN	**.740**	.524	**.585 ± .140**	.982	**.991**	**.986 ± .002**	**.861**	.758	**.786 ± .070**
IAN (all)	.443	.567	.496 ± .117	.983	.972	.977 ± .003	.713	**.770**	.737 ± .059

Table 5. Classification results of the compared methods for TwiMed-Twitter corpus.

Method	ADR			non-ADR			Macro		
	P	R	F	P	R	F	P	R	F
Feature-rich SVM	.752	.810	.778 ± .047	.779	.707	.739 ± .054	.766	.758	.758 ± .049
IAN	**.836**	**.813**	**.824 ± .042**	**.802**	**.825**	**.813 ± .036**	**.819**	**.819**	**.819 ± .039**
IAN (all)	.740	.757	.745 ± .060	.738	.711	.721 ± .038	.739	.734	.733 ± .046

Table 6. Classification results of the compared methods for TwiMed-PubMed corpus.

Method	ADR			non-ADR			Macro		
	P	R	F	P	R	F	P	R	F
Feature-rich SVM	.799	.681	.728±.100	.925	.955	.939 ± .017	.862	.818	.834 ± .054
IAN	**.878**	**.738**	**.792 ± .016**	**.935**	**.977**	**.956 ± .105**	**.907**	**.857**	**.874 ± .059**
IAN (all)	.660	.614	.633 ± .154	.905	.925	.915 ± .024	.783	.770	.774 ± .080

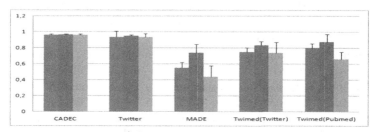

a) Precision performance.

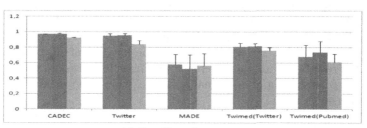

b) Recall performance.

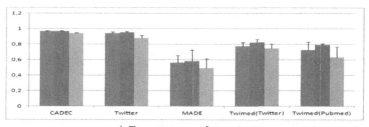

c) F-measure performance.

Fig. 2. Classification results for ADR class: (a) Precision (b) Recall (c) F-measure for three methods: blue - Feature-reach SVM, red - IAN, green - IAN (all). (Color figure online)

vocabulary cutoff of 10, Continuous Bag of Words model. Coverage statistics of word embedding model vocabulary: CADEC - 93.5%, Twitter - 80.4%, MADE - 62.5%, TwiMed-Twitter - 81.2%, TwiMed-Pubmed - 76.4%. For the words out of vocabulary the representations were uniformly sampled from the range of embedding weights. We used a maximum of 15 epochs to train IAN on each dataset, the batch size of 128, number of hidden units for LSTM layer 300, the learning rate of 0.01, l2 regularization of 0.001, dropout 0.5. We applied the implementation of the model from this repository[2].

All models were evaluated by 5-fold cross-validation. We computed averaged recall (R), precision (P) and F_1-measures (F) for classes ADR and non-ADR separately and then macro-average of these values for both classes. In the first set of experiments, IAN was evaluated on each dataset separately. In the second set of experiments, we joined all training subsets into a single dataset for training IAN and evaluated the model on 5 testing subsets. We mark the model trained on all five subsets as IAN (all). Tables 2, 3, 4, 5 and 6 present the results of experiments for each corpora.

The results show that the IAN outperformed baseline model in terms of macro-averaged measures for both ADR and non-ADR classes and for two classes. The most significant increase of macro average F-measure was obtained on Twitter (4.5%), Twimed-Pubmed (6.1%) and Twimed-Twitter (4%) corpora. SVM outperformed IAN only in term of recall for ADR class (5.8%) and precision for non-ADR class (0.02%) for MADE dataset. The most significant increase of macro average F-measure for ADR class detection was obtained on Twimed-Twitter (4.6%) and Twimed-Pubmed (6.4%) corpora.

IAN (all) outperformed IAN only in terms of recall for ADR class on MADE dataset. We assume that the combination of training data of all corpora did not increase the results due to the difference in vocabulary used in different data sets. MADE consists of a more formal language, while the rest of the case uses a simpler language model.

To sum up, the IAN approach shows the increase of results for all corpora. The most insignificant improvements were achieved for the CADEC corpora. Figure 2 presents diagrams for each metric: precision, recall, F-measure for all corpora and methods for ADR class.

6 Conclusion

In this paper, we explored the potential of IAN to the task of ADR classification. We tested the proposed approach on five benchmark corpora of user reviews, tweets, scientific articles and electronic health records. We compare results of this model with the previous state-of-the-art approach. Our experiments showed that IAN increased the results of F_1-measure for ADR class detection and macro-averaged measures for two classes. We also found out that combination of all data sets for training IAN do not increase the quality of classification. In the future we plan to compare IAN model with other neural network models for aspect based sentiment analysis for the task of ADR classification.

Acknowledgments. This work was supported by the Russian Science Foundation Grant No. 18-11-00284. The authors are grateful to Elena Tutubalina for useful discussions about this study.

---

[2] https://github.com/songyouwei/ABSA-PyTorch.

References

1. Alimova, I., Tutubalina, E.: Automated detection of adverse drug reactions from social media posts with machine learning. In: van der Aalst, W. (ed.) AIST 2017. LNCS, vol. 10716, pp. 3–15. Springer, Cham (2018). https://doi.org/10.1007/978-3-319-73013-4_1

2. Alvaro, N., Miyao, Y., Collier, N.: TwiMed: Twitter and PubMed comparable corpus of drugs, diseases, symptoms, and their relations. JMIR Public Health Surveill. **3**(2), e24 (2017)

3. Aramaki, E., et al.: Extraction of adverse drug effects from clinical records. In: MedInfo, pp. 739–743 (2010)

4. Baccianella, S., Esuli, A., Sebastiani, F.: SentiWordNet 3.0: an enhanced lexical resource for sentiment analysis and opinion mining. In: LREC, vol. 10, pp. 2200–2204 (2010)

5. Benton, A., et al.: Identifying potential adverse effects using the web: a new approach to medical hypothesis generation. J. Biomed. Inform. **44**(6), 989–996 (2011)

6. Bian, J., Topaloglu, U., Yu, F.: Towards large-scale Twitter mining for drug-related adverse events. In: Proceedings of the 2012 International Workshop on Smart Health and Wellbeing, pp. 25–32. ACM (2012)

7. Biyani, P., et al.: Co-training over domain-independent and domain-dependent features for sentiment analysis of an online cancer support community. In: Proceedings of the 2013 IEEE/ACM International Conference on Advances in Social Networks Analysis and Mining, pp. 413–417. ACM (2013)

8. Cambria, E., Benson, T., Eckl, C., Hussain, A.: Sentic PROMs: application of sentic computing to the development of a novel unified framework for measuring health-care quality. Expert. Syst. Appl. **39**(12), 10533–10543 (2012)

9. Chowdhury, S., Zhang, C., Yu, P.S.: Multi-task pharmacovigilance mining from social media posts. arXiv preprint arXiv:1801.06294 (2018)

10. Classen, D.C., Pestotnik, S.L., Evans, R.S., Lloyd, J.F., Burke, J.P.: Adverse drug events in hospitalized patients: excess length of stay, extra costs, and attributable mortality. JAMA **277**(4), 301–306 (1997)

11. Dai, H.J., Touray, M., Jonnagaddala, J., Syed-Abdul, S.: Feature engineering for recognizing adverse drug reactions from Twitter posts. Information **7**(2), 27 (2016)

12. Demner-Fushman, D., Chapman, W.W., McDonald, C.J.: What can natural language processing do for clinical decision support? J. Biomed. Inform. **42**(5), 760–772 (2009)

13. Denecke, K., Dolog, P., Smrz, P.: Making use of social media data in public health. In: Proceedings of the 21st International Conference on World Wide Web, pp. 243–246. ACM (2012)

14. Gupta, S., Gupta, M., Varma, V., Pawar, S., Ramrakhiyani, N., Palshikar, G.K.: Co-training for extraction of adverse drug reaction mentions from tweets. In: Pasi, G., Piwowarski, B., Azzopardi, L., Hanbury, A. (eds.) ECIR 2018. LNCS, vol. 10772, pp. 556–562. Springer, Cham (2018). https://doi.org/10.1007/978-3-319-76941-7_44

15. Harpaz, R., et al.: Text mining for adverse drug events: the promise, challenges, and state of the art. Drug Saf. **37**(10), 777–790 (2014)

16. Hou, Y., Li, X., Wu, G., Ye, X.: National ADR monitoring system in China. Drug Saf. **39**(11), 1043–1051 (2016). https://doi.org/10.1007/s40264-016-0446-5

17. Hu, M., Liu, B.: Mining and summarizing customer reviews. In: Proceedings of the Tenth ACM SIGKDD International Conference on Knowledge Discovery and Data Mining, pp. 168–177. ACM (2004)

18. Huang, B., Ou, Y., Carley, K.M.: Aspect level sentiment classification with attention-over-attention neural networks. arXiv preprint arXiv:1804.06536 (2018)
19. Huynh, T., He, Y., Willis, A., Rüger, S.: Adverse drug reaction classification with deep neural networks. In: Proceedings of COLING 2016, the 26th International Conference on Computational Linguistics: Technical papers, pp. 877–887 (2016)
20. Ivanov, V., Tutubalina, E., Mingazov, N., Alimova, I.: Extracting aspects, sentiment and categories of aspects in user reviews about restaurants and cars. In: Proceedings of International Conference Dialog, vol. 2, pp. 22–34 (2015)
21. Karimi, S., Metke-Jimenez, A., Kemp, M., Wang, C.: Cadec: a corpus of adverse drug event annotations. J. Biomed. Inform. **55**, 73–81 (2015)
22. Kiritchenko, S., Mohammad, S.M., Morin, J., de Bruijn, B.: NRC-Canada at SMM4H shared task: classifying tweets mentioning adverse drug reactions and medication intake. arXiv preprint arXiv:1805.04558 (2018)
23. Korkontzelos, I., Nikfarjam, A., Shardlow, M., Sarker, A., Ananiadou, S., Gonzalez, G.H.: Analysis of the effect of sentiment analysis on extracting adverse drug reactions from tweets and forum posts. J. Biomed. Inform. **62**, 148–158 (2016)
24. Lai, S., Xu, L., Liu, K., Zhao, J.: Recurrent convolutional neural networks for text classification. In: AAAI, vol. 333, pp. 2267–2273 (2015)
25. Lazarou, J., Pomeranz, B.H., Corey, P.N.: Incidence of adverse drug reactions in hospitalized patients: a meta-analysis of prospective studies. JAMA **279**(15), 1200–1205 (1998)
26. Li, H., Min, M.R., Ge, Y., Kadav, A.: A context-aware attention network for interactive question answering. In: Proceedings of the 23rd ACM SIGKDD International Conference on Knowledge Discovery and Data Mining, pp. 927–935. ACM (2017)
27. Liu, F., Yu, H., Jagannatha, A., Liu, W.: NLP challenges for detecting medication and adverse drug events from electronic health records (MADE1.0) (2018). https://bio-nlp.org/index.php/projects/39-nlp-challenges
28. Liu, Q., Zhang, H., Zeng, Y., Huang, Z., Wu, Z.: Content attention model for aspect based sentiment analysis. In: Proceedings of the 2018 World Wide Web Conference on World Wide Web, pp. 1023–1032. International World Wide Web Conferences Steering Committee (2018)
29. Liu, X., Chen, H.: AZDrugMiner: an information extraction system for mining patient-reported adverse drug events in online patient forums. In: Zeng, D. (ed.) ICSH 2013. LNCS, vol. 8040, pp. 134–150. Springer, Heidelberg (2013). https://doi.org/10.1007/978-3-642-39844-5_16
30. Ma, D., Li, S., Zhang, X., Wang, H.: Interactive attention networks for aspect-level sentiment classification. arXiv preprint arXiv:1709.00893 (2017)
31. Marelli, M., Bentivogli, L., Baroni, M., Bernardi, R., Menini, S., Zamparelli, R.: SemEval-2014 task 1: evaluation of compositional distributional semantic models on full sentences through semantic relatedness and textual entailment. In: Proceedings of the 8th International Workshop on Semantic Evaluation (SemEval 2014), pp. 1–8 (2014)
32. Melzi, S., Abdaoui, A., Azé, J., Bringay, S., Poncelet, P., Galtier, F.: Patient's rationale: patient knowledge retrieval from health forums. In: eTELEMED: eHealth, Telemedicine, and Social Medicine (2014)
33. Meng, F., Lu, Z., Li, H., Liu, Q.: Interactive attention for neural machine translation. arXiv preprint arXiv:1610.05011 (2016)
34. Meystre, S., Haug, P.J.: Natural language processing to extract medical problems from electronic clinical documents: performance evaluation. J. Biomed. Inform. **39**(6), 589–599 (2006)

35. Miftahutdinov, Z., Tutubalina, E., Tropsha, A.: Identifying disease-related expressions in reviews using conditional random fields. In: Computational Linguistics and Intellectual Technologies: Papers from the Annual conference "Dialogue", vol. 1, no. 16, pp. 155–166 (2017). http://www.dialog-21.ru/media/3932/miftahutdinovzshetal.pdf

36. Miranda, D.S.: Automated detection of adverse drug reactions in the biomedical literature using convolutional neural networks and biomedical word embeddings. arXiv preprint arXiv:1804.09148 (2018)

37. Na, J.-C., Kyaing, W.Y.M., Khoo, C.S.G., Foo, S., Chang, Y.-K., Theng, Y.-L.: Sentiment classification of drug reviews using a rule-based linguistic approach. In: Chen, H.-H., Chowdhury, G. (eds.) ICADL 2012. LNCS, vol. 7634, pp. 189–198. Springer, Heidelberg (2012). https://doi.org/10.1007/978-3-642-34752-8_25

38. Nikfarjam, A., Gonzalez, G.H.: Pattern mining for extraction of mentions of adverse drug reactions from user comments. In: AMIA Annual Symposium Proceedings, vol. 2011, p. 1019. American Medical Informatics Association (2011)

39. Nikfarjam, A., Sarker, A., O'Connor, K., Ginn, R., Gonzalez, G.: Pharmacovigilance from social media: mining adverse drug reaction mentions using sequence labeling with word embedding cluster features. J. Am. Med. Inform. Assoc. **22**(3), 671–681 (2015)

40. Niu, Y., Zhu, X., Li, J., Hirst, G.: Analysis of polarity information in medical text. In: AMIA Annual Symposium Proceedings, vol. 2005, p. 570. American Medical Informatics Association (2005)

41. Oliveira, J.L., et al.: The EU-ADR web platform: delivering advanced pharmacovigilance tools. Pharmacoepidemiol. Drug Saf. **22**(5), 459–467 (2013)

42. de la Peña, S., Segura-Bedmar, I., Martínez, P., Martínez, J.L.: ADRSpanishTool: a tool for extracting adverse drug reactions and indications. Procesamiento del Lenguaje Natural **53**, 177–180 (2014)

43. Pirmohamed, M., et al.: Adverse drug reactions as cause of admission to hospital: prospective analysis of 18 820 patients. BMJ **329**(7456), 15–19 (2004)

44. Raju, G.S., et al.: Natural language processing as an alternative to manual reporting of colonoscopy quality metrics. Gastrointest. Endosc. **82**(3), 512–519 (2015)

45. Rastegar-Mojarad, M., Elayavilli, R.K., Yu, Y., Liu, H.: Detecting signals in noisy data-can ensemble classifiers help identify adverse drug reaction in tweets. In: Proceedings of the Social Media Mining Shared Task Workshop at the Pacific Symposium on Biocomputing (2016)

46. Rodrigues, R.G., das Dores, R.M., Camilo-Junior, C.G., Rosa, T.C.: SentiHealth-cancer: a sentiment analysis tool to help detecting mood of patients in online social networks. Int. J. Med. Inform. **85**(1), 80–95 (2016)

47. Rusnachenko, N., Loukachevitch, N.: Using convolutional neural networks for sentiment attitude extraction from analytical texts. In: Proceedings of CEUR Workshop, CLLS-2018 Conference. CEUR-WS.org (2018)

48. Salas-Zárate, M.D.P., Medina-Moreira, J., Lagos-Ortiz, K., Luna-Aveiga, H., Rodríguez-García, M.Á., Valencia-García, R.: Sentiment analysis on tweets about diabetes: an aspect-level approach. Comput. Math. Methods Med. **2017**, 9 (2017)

49. Sarker, A., et al.: Utilizing social media data for pharmacovigilance: a review. J. Biomed. Inform. **54**, 202–212 (2015)

50. Sarker, A., Gonzalez, G.: Portable automatic text classification for adverse drug reaction detection via multi-corpus training. J. Biomed. Inform. **53**, 196–207 (2015)

51. Sboev, A., Litvinova, T., Voronina, I., Gudovskikh, D., Rybka, R.: Deep learning network models to categorize texts according to author's gender and to identify text sentiment. In: 2016 International Conference on Computational Science and Computational Intelligence (CSCI), pp. 1101–1106. IEEE (2016)

52. Serrano-Guerrero, J., Olivas, J.A., Romero, F.P., Herrera-Viedma, E.: Sentiment analysis: a review and comparative analysis of web services. Inf. Sci. **311**, 18–38 (2015)

53. Shareef, S., Naidu, C., Raikar, S.R., Rao, Y.V., Devika, U.: Development, implementation, and analysis of adverse drug reaction monitoring system in a rural tertiary care teaching hospital in Narketpally, Telangana. Int. J. Basic Clin. Pharmacol. **4**(4), 757–760 (2017)

54. Singh, P., Agrawal, M., Hishikar, R., Joshi, U., Maheshwari, B., Halwai, A.: Adverse drug reactions at adverse drug reaction monitoring center in Raipur: analysis of spontaneous reports during 1 year. Indian J. Pharmacol. **49**(6), 432 (2017)

55. Sokolova, M., Matwin, S., Jafer, Y., Schramm, D.: How Joe and Jane tweet about their health: mining for personal health information on Twitter. In: Proceedings of the International Conference Recent Advances in Natural Language Processing, RANLP 2013, pp. 626–632 (2013)

56. Solovyev, V., Ivanov, V.: Dictionary-based problem phrase extraction from user reviews. In: Sojka, P., Horák, A., Kopeček, I., Pala, K. (eds.) TSD 2014. LNCS (LNAI), vol. 8655, pp. 225–232. Springer, Cham (2014). https://doi.org/10.1007/978-3-319-10816-2_28

57. Wang, B., Liu, K., Zhao, J.: Inner attention based recurrent neural networks for answer selection. In: Proceedings of the 54th Annual Meeting of the Association for Computational Linguistics (Volume 1: Long Papers), vol. 1, pp. 1288–1297 (2016)

58. Wilson, T., Wiebe, J., Hoffmann, P.: Recognizing contextual polarity in phrase-level sentiment analysis. In: Proceedings of the Conference on Human Language Technology and Empirical Methods in Natural Language Processing, pp. 347–354. Association for Computational Linguistics (2005)

59. Xu, H., Anderson, K., Grann, V.R., Friedman, C.: Facilitating cancer research using natural language processing of pathology reports. In: Studies in Health Technology and Informatics (2004)

60. Yang, M., Qu, Q., Chen, X., Guo, C., Shen, Y., Lei, K.: Feature-enhanced attention network for target-dependent sentiment classification. Neurocomputing **307**, 91–97 (2018)

61. Yang, Z., Yang, D., Dyer, C., He, X., Smola, A., Hovy, E.: Hierarchical attention networks for document classification. In: Proceedings of the 2016 Conference of the North American Chapter of the Association for Computational Linguistics: Human Language Technologies, pp. 1480–1489 (2016)

Modeling Propaganda Battle: Decision-Making, Homophily, and Echo Chambers

Alexander Petrov[1] and Olga Proncheva[1,2(✉)]

[1] Keldysh Institute of Applied Mathematics, Moscow, Russia
petrov.alexander.p@yandex.ru
[2] Moscow Institute of Physics and Technology, Dolgoprudny, Russia
olga.proncheva@gmail.com

Abstract. Studies concerning social patterns that appear as a result of propaganda and rumors generally tend to neglect considerations of the behavior of individuals that constitute these patterns. This places obvious limitations upon the scope of research. We propose a dynamical model for the mechanics of the processes of polarization and formation of echo chambers. This model is based on the Rashevsky neurological scheme of decision-making.

Keywords: Social homophily · Polarization · Selective exposure
Echo chambers · Decision-making · Rashevsky's neurological scheme
Rumors

1 Introduction

There used to be a time when deep theories and spectacular insights populated the field of propaganda research, including agenda-setting theory [1] or books written by Chomsky [2,3], which are now considered as classics. However, a Web 2.0 type of communication did not exist at that time. Propaganda was viewed as a phenomenon in which individuals merely received messages from mass-media, without or hardly circulating any message among other individuals. Accordingly, theoretical and empirical work was focused on mass-media and its influence, motives, biases, and behavior.

The advent of online social networks has created a tremendous difference both in communication and the research concerned with it. There was a realization that the diffusion of information through interpersonal communications has become immensely powerful, with some individuals being able to broadcast widely and some pieces of information going viral. The empirical literature pertaining to the diffusion of information through Twitter and Facebook surged, and the new and increasingly important science of information retrieval emerged. Even artificial societies as models of reality have come into use. We can

Supported by the Russian Foundation for Basic Research (project 17-01-00390 A).

D. Ustalov et al. (Eds.): AINL 2018, CCIS 930, pp. 197–209, 2018.
https://doi.org/10.1007/978-3-030-01204-5_19

see numerous and diverse advances in empirical studies, which have advanced beyond the scope of development of theories of propaganda.

However, there is a paradox with this burst of research. Despite all the prevalent awareness regarding the increased role of individuals, there are no models that focus on the behavior of individuals in propaganda battles. Furthermore, the related important question is whether the battle creates a sustainable change in individuals. Suppose that a controversial issue enters the public agenda. People discuss it, they tend to get harsh with each other, and researchers witness more polarization than before. Afterwards, this controversial issue is no longer on the agenda. Yet, what is the longer term impact on the individuals? Do they stay highly polarized or relax back to their initial state? It is important due to the fact that there will be more battles in the future, and nobody wants their society to be torn apart.

In order to address this question, we look at individual behavior and not solely at the social structure. We opine that the promising research effort would be to borrow some ideas from another subfield of propaganda research. That subfield examines the kind of propaganda that intentionally seeks radicalization of individuals into violent extremism. The useful invention here is the approach that focuses on the evolution of an individual's mindset, such as Borum's four-stage model [4,5] "Grievance (It's not fair)" - "Injustice (It's not right)" - "Target attribution (It's your fault)" - "Devaluation (You're evil)". Some other models of the same type describe the process of radicalization as a gradual shift of an individual from lower to higher levels of radicalization [6–8]. We apply this idea to the topic of propaganda battles by assuming that an individual's attitude can be represented as a point on a spectrum ranging from strong support of the Left party to strong support of the Right party. During the propaganda battle, an individual can potentially undergo any shift along the spectrum. We also employ the idea that the central point of the analysis is the way in which stimuli affect an individual's choice, and we employ Rashevsky's neurological scheme [9] to formalize this idea.

This collection of ideas provides us with the decision-making mechanism that allows a new perspective on social phenomena, such as polarization and echo chambers. These echo chambers are related to the question posed above: Do people stay highly polarized after the propaganda battle? Alternatively, do they relax back to the relatively moderate political views they had before the initiation of the battle? The answer proposed here is that they relax back, unless an echo chamber has been formed. In other words, they stay radicalized if the number of these radicalized individuals is high enough, and they keep perpetuating radical views within their group and prevent one another from relaxing.

Accordingly, this paper, technically speaking, is mostly concerned with modeling the dynamical process of the formation of an echo chamber.

In the next section, we set forth our approach in layman's terms. In Sect. 3, we formulate the basic mathematical model that illustrates the approach under simplifying assumptions. Section 4 introduces and analyses the workhorse model.

2   The Approach in Layman's Terms

2.1   Propaganda Battle

Consider a population in which two parties Left (L) and Right (R) are engaged in a propaganda battle. They broadcast their messages via affiliated mass-media. At any given moment, each member of the population approves one of the parties and disapproves of the other. They communicate with one another and share their approvals. Over the course of time, an individual may shift their support to the other party under the influence of belligerent parties' broadcasting and the opinions of other individuals. Theoretically, they may switch their partisanship back and forth an unlimited number of times. Therefore, the number of supporters for each party varies over time. This process is called a propaganda battle.

The precursors to the models of information wars are single-rumor models. The first of them was proposed by Daley and Kendall [10] as long ago as 1964. They introduced the first of the single-rumor models. Another early model with similar foundation was introduced by Maki and Thompson [11]. In the most general terms, these models assume that we have a closed group of individuals, and at every point in time some of them have a certain valuable piece of information and spread it among other individuals. Thus, there is a spread of single rumor (in this field, the term "rumor" does not necessarily mean that the information is unverified, and is used as an umbrella term for rumors, gossips and urban legends, which are, rigorously speaking, quite different things [12]). Today there is extensive literature on the modelling of rumors, which has been developed mainly as a branch of mathematics in distinct separation from the social sciences. Most of it develops on the Daley-Kendall and Maki-Thompson approaches. For example, the model with latent constant recruitment where the total population varied over time was studied in [13], and the model with several groups of spreaders was considered in [14]. A model considering rumor transmission with incubation was introduced in [15]. This model considers constant recruitment and implies the possibility of contagion during both latent and infected periods. The full range of possible demographic events (birth, death, emigration and immigration of individuals) was included in the rumor model in [16]. Some papers are aimed at mathematical generalization of the models [17,18]. These are only some of the vast number of papers that develop the Daley-Kendall and Maki-Thompson approaches.

However, there is a drawback to these approaches. They both lead to the grotesque conclusion that all rumors (that have ever been in any historical age and in any society) encompass the same proportion of the population at the end of its circulation. If the initial number of spreaders is 1, then it follows from both Daley-Kendall and Maki-Thompson models that, whatever the rumor is, the final $(t \to \infty)$ proportion of persons who have never learned the rumor is approximately 20.3% of the whole population. It was also shown for the Maki-Thompson model that if the initial number of spreaders is close enough to the whole population, then the number of persons who have never learned the rumor is approximately $1/e$ of their initial number [19,20].

This grotesque conclusion appears due to the notion that rumors stop being propagated due to communication alone. Here, the mechanism of suppressing a rumor is that when two spreaders interact, they both (in Daley-Kendall) or one of them (in Maki-Thompson) come to the idea that this rumor is not news anymore and is no longer worth spreading further. As a result, it appears in that mechanics that at the end of the process, all the spreaders have given up spreading the rumor due to communication with one another, while some individuals still remain ignorant of the entire rumor.

Under this notion, the property of a specific rumor to be exciting is irrelevant to its fading. This notion mismatches the reality of the situation, in which some rumors cover almost all the adult population of a country, whereas some other rumors stop being propagated as early as at the second or third spreader, i.e., they actually even fail to become rumors in the most mundane meaning of the word. In real life, people simply do not share humdrum information with interlocutors, or they may quickly lose their interest in such information. We can safely say that the Daley-Kendall and Maki-Thompson models fail to distinguish between propagating viral and humdrum pieces of information.

However, many models of competing rumors inherit the mechanics of Daley-Kendall model together with this grotesque feature. The earliest model of this kind was proposed by Osei and Thomson [21], who considered the dissemination of two competing rumors, wherein the second was seen as stronger. In other words, when the spreader of the first rumor meets the spreader of the second rumor, the spreader of the first rumor adopts the second rumor. The idea underlying this model is that the first rumor is fake, and that the second rumor possesses highly convincing evidence of the falseness. There were also some later models of competing rumors, such as in [22]. These papers also develop Daley-Kendall and Maki-Thompson approaches and their mechanics. Quite a different model of information warfare was suggested by Mikhailov and Marevtseva [23]. It had two chief novelties. The first is that mass-media was introduced as a source of information. In other words, it did not consider the dissemination of information via rumors alone, but rather via both rumors and mass-media. The second novelty is that once attached to a party, an individual was supposed to remain with that party forever. Therefore, it was possible to pose the question regarding which party will eventually control the majority of the population.

Other approaches to mathematical modelling of rumors and propaganda wars include emphasis on social networks, and agent-based and game theory-based models [24–28].

Recent years have been remarkable for the growing research interest in issues related to propaganda (with most of the research being empirical). Table 1 shows the annual number of papers with the exact phrase "fake news" (Google Scholar, retrieved March 25, 2018). The use of this phrase was apparently triggered by the US Presidential campaign (in which pro-Trump media and bloggers popularized this phrase) and the subsequent allegations of Russian interference in the elections (which were pushed into agenda by anti-Trump media and bloggers). With this backdrop, the topic of propaganda research seems to be entering its own Golden Age.

Table 1. The annual numbers of research papers with the exact phrase "fake news".

2003	2004	2005	2006	2007	2008	2009	2010	2011	2012	2013	2014	2015	2016	2017
196	220	255	300	339	385	351	395	556	520	595	629	612	924	7380

2.2 Decision-Making

The decision-making mechanism utilized in this study is based on Rashevsky's neurological scheme [9]. The outcome of this mechanism is individuals' support of the Left or the Right party, which we refer to as their manifest political position. It manifests their latent political position, which, mathematically speaking, is a continuous scalar function of time. This can be imagined as an axis of latent political positions where positive values refer to the support of the Right party (manifest position $= R$) and negative values refer to the support of the Left party (manifest position $= L$). The greater the absolute value is, the stauncher a supporter of the party the individual is. Over the course of time, their latent position moves along the axis, and if it crosses the zero point, the individual begins supporting the other party.

In its turn, the latent position of the individual is a sum of their attitude φ and the dynamical component $\psi(t)$, which is the same for all members of the population (it is so in our basic model, Sect. 3, but it is slightly more complicated in the workhorse model, Sect. 4).

Attitude is a term used to describe a person's predisposition towards a given object, a settled inclination to assess this object in a certain way, and the way to feel about it and to act in the context of this object. For instance, some Americans bear more or less extensive positive attitudes towards Donald Trump, while others have negative attitudes. When two Americans are reading the same critical newspaper article about Mr. Trump, one of the readers may feel angry with Trump, whereas the other may feel angry with the author of the article. This is the manner in which their attitudes affect their perceptions of the text. In our model, attitude is bipolar. In other words, it is a person's predisposition to one of the two parties in relation to the other party.

Attitude generally depends on a person's social experience and status. It is important for the approach that this attitude is permanent or at least long-lasting (in contrast to mood). In our model, attitudes of all members of the population are supposed to be developed before the commencement of the propaganda battle and to remain constant over the whole course of it. The distribution of attitudes among individuals in a polarized population is a two-peaked function $n(\varphi)$.

The dynamical component of the latent political position changes over time during the battle due to informational factors - these are propaganda broadcast and mouth-to-mouth exchanges. Both factors are important. This follows, for instance, from the study of the Rwandian genocide [29], which found that the violence was more extreme in villages covered by the broadcast of RTLM radio (which encouraged ethnic cleansing) and also in other villages connected with them via interpersonal communications of inhabitants.

Therefore, attitude is constant over time, but varies across individuals; conversely, the dynamical component varies over time, but is constant across individuals. Our approach may also be said to emphasize two time-scales. Long-term factors shape attitude, which operates as a background or context. Short-run factors influence the dynamical component, which varies and adds to this background.

The dynamics looks as follows. As previously explained, an individual supports Right party (Party R) if $\varphi + \psi(t) > 0$, and they support Left party (Party L) if $\varphi + \psi(t) < 0$. Therefore, the growth of $\psi(t)$ is in favor of Party R and vice versa. So, the competition between the parties may be considered a tug-of-war in which $\psi(t)$ represents the rope.

The main equation of the model is a differential equation for $\psi(t)$, in which the factors in favor of Party R contribute to the increase of $\psi(t)$ and the factors in favor of Party L contribute to the decrease of $\psi(t)$. This differential equation is based on psychological considerations [9]; however, it can be basically explained in sociological terms, which is done in Sect. 3.

Three notes should also be highlighted. The first note is that these variables cannot be measured directly. Yet, in applying models of this type to concretely defined problems, some combinations of their parameters can usually be estimated from observable data, such as in [30]. This makes the models usable and allows for some findings concerning real social phenomena.

The second note is that, in this paper, we adopt the simplifying assumption that the intensities of parties' broadcasting are constant values during the propaganda battle (though the battle may be followed by another battle with other intensities of propaganda).

The third note is that our model should not be confused with the median voter model from social choice theory. In public choice theory, any voter has their own preferences with no relation to political parties, and there may be any number of these parties. This class of models treats political positions as voters' statement, for example, "Military expenditure should be 3% of GDP". The voter casts their vote for the party whose statement is the closest to their own 3%. It is quite different in our model, where there are exactly two parties, and the individual's political position is their positioning with regard to and between them. Attitude may be viewed as their devotedness or attachment to the party (regardless of the reasons for supporting the party).

2.3   Homophily, Selective Exposure, Polarization, and Echo Chambers

The above subsection was concerned with the mechanism inside the individual, whereas the following are the outlines of some social patterns.

Similarity breeds connection. This succinct formula was introduced in the very influential paper [31] by McPherson, Smith-Lovin, and Cook (which has been cited more than 12000 times since 2001). This is the idea behind homophily, which is the preferential matching in many contexts. In our specific context, homophily is the tendency to communicate and exchange views with other individuals who are considerably similar. In our workhorse model, we consider a population that comprises two groups, as each individual prefers to communicate with other members of the same group.

Selective exposure means that individuals select the media they are exposed to. The key postulate regarding politically motivated selective exposure is that people prefer exposure to media that supports their own views and beliefs (see [32] for the discussion). There is also another kind of selective exposure, which is related to an individual's demographic and social status. In other words, a media outlet has been utilized by a specific demographic or/and social group that this outlet targets. For example, the target audience of "Russia 1" TV channel is the 45–65 age group, while the target audience of STS channel is the 10–45 age group. Some media may be utilized mostly by metropolitans, lower middle class individuals, women, blue collar workers, and so on. In our workhorse model, we use this socially motivated selective exposure by considering the two groups (which make up the whole population) that are differently exposed.

Polarization is a big topic in political science [33, 34]. In most of the experiments with our workhorse model, the attitudes of the members of both groups are such that these groups are predisposed towards different parties. Note that the growth of polarization here means that individuals become more attached to their parties, and their doubts and hesitations are reduced. (In public choice theory, polarization is quite different: parties and individuals on the right move further right and those on the left move further left).

An echo chamber is a social pattern wherein some people have beliefs or views that are distinctly different from those that are common within the broader population, and these people preferably communicate with one another, and through this, they manage to maintain, reinforce, or amplify these beliefs or views (an intriguing question for natural language processing scholars would be whether the population of an echo chamber develops its own "dialect" that is distinguishable from the mainstream political language of the nation). Echo chambers appear due to homophily, selective exposure, and polarization, as these factors provide reinforcement. Once formed, an echo chamber maintains itself through these factors. In our model, echo chambers are a type of stable solution where a minority of the population continues to support one of the parties despite the domination of the other party.

3   Basic Model

In this section, we present the simplest model of this kind. It was introduced in [35] and is shown here just for explanatory purposes. Our aim here is to explain the model of decision-making. There is neither homophily, nor selective exposure in this model.

An individual supports Party R if $\varphi + \psi(t) > 0$; therefore, the number of these supporters is given by

$$R(\psi(t)) = \int_{-\psi(t)}^{\infty} n(\varphi)\, d\varphi, \tag{1}$$

and the number of the supporters of Party L is

$$L(\psi(t)) = \int_{-\infty}^{-\psi(t)} n(\varphi)\, d\varphi, \tag{2}$$

where $n(\varphi)$ is the distribution of attitudes among individuals.

It was previously stated in Subsect. 2.2 that the main equation of the model is a differential equation for $\psi(t)$, in which the factors in favor of Party R contribute to the increase of $\psi(t)$ and the factors in favor of Party L contribute to its decrease. The whole factor in favor of Party R is $CR(\psi(t)) + b_R$. Here, b_R is the intensity of this party's media propaganda, $R(\psi(t))$ is the number of individuals arguing in its favor, and C is a positive constant describing the degree of activism in arguing. In the same way, the whole factor in favor of Party L is $CL(\psi(t)) + b_L$. The difference of these factors is the driving force for $\psi(t)$. The equation is

$$\frac{d\psi}{dt} = C[R(\psi) - L(\psi)] - b_R - b_L - a\psi, \tag{3}$$

where $a > 0$ and the term $-a\psi$ describes the relaxation. This means that in the case of a hypothetical removal of pro-R and pro-L factors, the dynamical factor $\psi(t)$ would gradually vanish and the latent position $\varphi + \psi(t)$ would tend to φ; that is, each individual would gradually relax back to their initial attitude.

After putting (1), (2) into (3), we get our basic model [35]

$$\frac{d\psi}{dt} = C\left[2\int_{-\psi(t)}^{\infty} n(\varphi) - N_0\right] - b_R - b_L - a\psi, \tag{4}$$

where N_0 stands for the entire number of individuals. Eq. (4) can also be derived from the apparently different approach, that is, from Rashevsky's neurological scheme [9]. Our way of obtaining this equation may be regarded as the sociological perspective, whereas Rashevsky's way is the psychological perspective, which is more difficult to comprehend but provides the founded meaning of the parameters.

The initial condition for ordinary differential Eq. (4) is

$$\int_{-\psi(0)}^{\infty} n(\varphi)\, d\varphi = R(0).$$

Clearly, this model oversimplifies the process in many ways. It makes no consideration of homophily, and all individuals are supposed to be equally exposed to the media. Therefore, in the next section, we develop a more comprehensive model that abandons these two oversimplifications.

4 Workhorse Model and the Findings

Our workhorse model has the following form (which is a generalization of the basic model (4)):

$$
\frac{d\psi_1}{dt} = C \left[\gamma \left(2 \int_{-\psi_1}^{\infty} n_1(\varphi) - N_1 \right) + (1-\gamma) \left(2 \int_{-\psi_2}^{\infty} n_2(\varphi) - N_2 \right) \right]
$$
$$
+ (b_{R1} - b_{L1}) - a\psi_1, \tag{5}
$$

$$
\frac{d\psi_2}{dt} = C \left[(1-\gamma) \left(2 \int_{-\psi_1}^{\infty} n_1(\varphi) - N_1 \right) + \gamma \left(2 \int_{-\psi_2}^{\infty} n_2(\varphi) - N_2 \right) \right]
$$
$$
+ (b_{R2} - b_{L2}) - a\psi_2. \tag{6}
$$

Here, the population comprises two groups; γ is homophily. For example, $\gamma = 1$ means each individual only communicates with members of their own group, and $\gamma = 0.5$ refers to the situation in which individuals have no preferences in communication. The indices 1 and 2 refer to the groups. For example, b_{R1} is the intensity of Party R's propaganda in the first group. The distributions of attitudes $n_1(\varphi)$, $n_2(\varphi)$ are Cauchy distributions

$$
f_i(\psi) = N_i \left(\pi \sigma \left[1 + (\psi - \psi_i)/\sigma \right] \right)^{-1} \tag{7}
$$

with location parameters $\psi_1 = -p$, $\psi_2 = p$ (here, we introduce polarization $p \geq 0$). Thus, the first group is generally predisposed towards Party L and the second group is predisposed towards Party R.

The numbers of party supporters in each group is given by formulae analogous to (1), (2); for example,

$$
R_1(\psi_1(t)) = \int_{-\psi_1(t)}^{\infty} n_1(\varphi)\, d\varphi \tag{8}
$$

is the number of supporters of Party R in the first group.

We studied the model analytically and numerically (by solving Eqs. (5), (6) with a variety of parameters) and obtained some findings that are not exactly counterintuitive but could hardly be assumed without resorting to mathematical modeling.

Analytical consideration shows that system (5), (6) can have 1 to 9 equilibrium points, depending on the number of intersections of the N-shaped and

S-shaped curves on the plane (ψ_1, ψ_2). Figure 1 relates to the most complex case of 9 fixed points. Here the central point is an unstable node, four points at the corners are stable nodes, and the remaining four points are saddles.

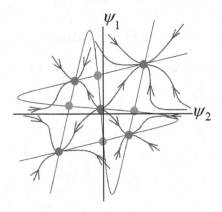

Fig. 1. One of the possible cases of the phase portrait of system (5), (6)

Each stable equilibrium point relates to certain values of $\psi_1 (t \to \infty)$, $\psi_2 (t \to \infty)$ and therefore to certain values of $R_i (t \to \infty)$, $L_i (t \to \infty)$, $i = 1, 2$. For example, for the upper right stable node we have $\psi_1 > 0$, $\psi_2 > 0$. Therefore $R_1 (t) > L_1 (t)$, $R_2 (t) > L_2 (t)$ is at $t \to \infty$, i.e. the Right party gets the majority in both groups. In this case, there is a small echo chamber comprising scanty supporters of the Left party from both groups.

Considering the lower right stable node in Fig. 1, we obtain a situation in which the Left party gets the majority in the first group (because $\psi_1 < 0$), and the Right party's majority in the second group is not so strong as in the previous case (because the lower right stable node is slightly to the left of the upper right stable node). If the groups are equal in size, each of the parties enjoys the support of approximately one-half of the population. However, if the second group is relatively small, then the majority of the population supports the Left party while there is an echo chamber constituted by the supporters of the Right party. This relatively small part of the population comprises the minority of the first group and the majority of the second group.

The remaining two stable nodes can be analyzed in the same way. Which one of the four equilibrium points is to be reached as $t \to \infty$, depends on the initial condition to system (5), (6).

Furthermore, a massive number of numerical experiments have been conducted by varying polarization p, homophily γ, and selectiveness of exposure ($b_{R1} - b_{L1}$ in comparison to $b_{R2} - b_{L2}$). The cases were considered not only when all three factors were present ($\gamma > 1/2$, $b_{R1} - b_{L1} \neq b_{R2} - b_{L2}$) but also when only one or two of them were present.

Our findings do not contradict the intuition regarding the idea that greater values of these factors facilitate the formation of echo chambers. Nonetheless, these findings depict a more complicated picture. We often saw in our experiments that a fixed set of parameters allows for two possibilities: a smaller or larger echo chamber may occur based on the initial condition (or, what is approximately the same for practical purposes, on the parties' preparations for battle).

For example, take $N_1 = 8$, $N_2 = 2$, $\sigma = 1$, $a = 0.8$, $C = 1$, $b_{R1} - b_{L1} = -0.4$, $b_{R2} - b_{L2} = 0.65$, $p = 0$. In other words, the groups are not polarized in attitudes; the first group tends to be exposed to the left-wing media and the second group tends to be exposed to the right-wing media. However, the first group is four times larger, and the intra-group communication is nine times more active than the communication between the groups. This weak inter-group communication makes us think that the first group would go to the Left party and the second group would go to the Right party. However, computations reveal that it is not so simple. In fact, there are two possibilities. One is that the Right party captures 3.5% of the first group and 18.3% of the second group, i.e., 6.5% of the whole population (a smaller echo chamber). The second possibility is that the Right party obtains 3.6% of the first group and 75.5% of the second group, i.e., 18.0% of the whole population (a large echo chamber). The formation of either of the two possible echo-chambers depends on the initial condition. Obviously, in a given real-life scenario, it would not be possible for the parameters of the model to be estimated very accurately. Nevertheless, this situation of two possible echo chambers must occur quite often in real life since, in the model, it appears under a wide variety of parameters.

The initial conditions (apparently, rare in real life) are also important in situations where the population's attitude is asymmetric in favor of, say, the Right party, but the initial condition is strongly favorable for the Left party. Sometimes, the initial condition is more powerful than the attitude, and vice versa.

There are also some less important findings. For example, if p and/or γ are great enough, there is only one stable stationary solution to the system (5), (6). In other words, in a highly polarized or separated society, a short-term boost in media propaganda would not result in any long-term outcome.

References

1. McCombs, M.E., Shaw, D.L.: The agenda-setting function of mass-media. Public Opin. Q. **36**(2), 176–187 (2016)
2. Chomsky, N.: Media Control: The Spectacular Achievements of Propaganda. Seven Stories Press, New York (1997)
3. Herman, E.S., Chomsky, N.: Manufacturing Consent: The Political Economy of the Mass-Media. Pantheon Books, New York (1988)
4. Borum, R.: Understanding the terrorist mindset. FBI Law Enforc. Bull. **72**(7), 7–10 (2003)
5. Borum, R.: Radicalization into violent extremism II: a review of conceptual models and empirical research. J. Strateg. Secur. **4**(4), 37–62 (2011)
6. Moghaddam, F.M.: The staircase to terrorism: a psychological exploration. Am. Psychol. **60**, 161–169 (2005)

7. Silber, M.D., Bhatt, A.: Radicalization in the West: The Homegrown Threat. NYPD Intelligence Division, New York Police Department, City of New York (2007)
8. Precht, T.: Home grown terrorism and Islamist radicalization in Europe: from conversion to terrorism. Danish Ministry of Defense (2007)
9. Rashevsky, N.: Mathematical Biophysics: Physico-Mathematical Foundations of Biology. Chicago Press, Univ. of Chicago (1938). University of Chicago
10. Daley, D.J., Kendall, D.G.: Stochastic rumors. J. Inst. Math. Appl. **1**, 42–55 (1964)
11. Maki, D.P., Thompson, M.: Mathematical Models and Applications. Prentice-Hall, Englewood Cliffs (1973)
12. DiFonzo, N., Bordia, P.: Rumor, Gossip and urban legends. Diogenes **54**(1), 19–35 (2007)
13. Chen, G., Shen, H., Ye, T., Chen, G., Kerr, N.: A kinetic model for the spread of rumor in emergencies. Discrete Dyn. Nat. Soc. **2013**, 1–8 (2013)
14. Isea, R., Mayo-García, R.: Mathematical analysis of the spreading of a rumor among different subgroups of spreaders. Pure Appl. Math. Lett. **2015**, 50–54 (2015)
15. Huo, L., Huang, P., Guo, C.: Analyzing the dynamics of a rumor transmission model with incubation. Discrete Dyn. Nat. Soc. **2012**, 1–21 (2012). Article ID 328151
16. Kawachi, K.: Deterministic models for rumor transmission. Nonlinear Anal. Real World Appl. **9**(5), 1989–2028 (2008)
17. Pearce, C.E.: The exact solution of the general stochastic rumour. Math. Comput. Model. Int. J. **31**(10–12), 289–298 (2000)
18. Dickinson, R.E., Pearce, C.E.M.: Rumours, epidemics, and processes of mass action: synthesis and analysis. Math. Comput. Model. **38**(11–13), 1157–1167 (2003)
19. Belen, S.: The behaviour of stochastic rumours. Ph.D. Thesis, The University of Adelaide (2008)
20. Belen, S., Pearce, C.E.M.: Rumours with general initial conditions. ANZIAM J. **4**, 393–400 (2004)
21. Osei, G.K., Thompson, J.W.: The supersession of one rumour by another. J. Appl. Prob. **14**(1), 127–134 (1977)
22. Escalante, R., Odehnal, M.: A deterministic mathematical model for the spread of two rumors. ArXiv preprint arXiv:1709.01726 (2017)
23. Mikhailov, A.P., Marevtseva, N.A.: Models of information warfare. Math. Model. **23**(10), 19–32 (2011)
24. Kaligotla, C., Yucesan, E., Chick, S.E.: An agent based model of spread of competing rumors through online interactions on social media. In: 9th International Proceedings of the 2015 Winter Simulation Conference, pp. 3985–3996 (2015)
25. Breer, V.V., Novikov, D.A., Rogatkin, A.D.: Mob Control: Models of Threshold Collective Behavior. SSDC, vol. 85. Springer, Cham (2017). https://doi.org/10.1007/978-3-319-51865-7
26. Gubanov, D.A., Novikov, D.A., Chkhartishvili, A.G.: Informational influence and informational control models in social networks. Autom. Remote Control **72**(7), 1557–1567 (2011)
27. Doerr, B., Fouz, M., Friedrich, T.: Why rumors spread so quickly in social networks. Commun. ACM **55**(6), 70–75 (2012)
28. Zhang, Y., Tang, C., Weigang, L.: Cooperative and competitive dynamics model for information propagation in online social networks. J. Appl. Math. **2014**, 1–12 (2014). Article ID 610382
29. Yanagizawa-Drott, D.: Propaganda and conflict: evidence from the rwandan genocide. Q. J. Econ. **129**(4), 1947–1994 (2014)

30. Mikhailov, A., Petrov, A., Pronchev, G., Proncheva, O.: Modeling a decrease in public attention to a past one-time political event. Dokl. Math. **97**(3), 247–249 (2018)
31. McPherson, M., Smith-Lovin, L., Cook, J.M.: Birds of a feather: homophily in social networks. Ann. Rev. Sociol. **27**(1), 415–444 (2001)
32. Garrett, R.K.: Politically motivated reinforcement seeking: reframing the selective exposure debate. J. Commun. **59**(4), 676–699 (2009)
33. Fiorina, M.P., Abrams, S.J.: Political polarization in the American public. Ann. Rev. Polit. Sci. **26**(10), 1531–1542 (2015)
34. Barbera, P., Jost, J.T., Nagler, J., Tucker, J.A., Bonneau, R.: Tweeting from left to right: is online political communication more than an echo chamber? Psychol. Sci. **26**(10), 1531–1542 (2015)
35. Petrov, A.P., Maslov, A.I., Tsaplin, N.A.: Modeling position selection by individuals during information warfare in society. Math. Models Comput. Simul. **8**(4), 401–408 (2016)

Author Index